# A STUDY ON THE EFFECT OF PLASMA MODIFICATION ON THE COMFORT PROPERTIES OF POLYESTER/COTTON BLEND FABRIC

**BAYE BERHANU YILMA**

**ETHIOPIAN INSTITUTE OF TEXTILE AND FASHION TECHNOLOGY [EiTEX]**
**BAHIR DAR UNIVERSITY**

**2021**

# A STUDY ON THE EFFECT OF PLASMA MODIFICATION ON THE COMFORT PROPERTIES OF POLYESTER/COTTON BLEND FABRIC

**By**

**Baye Berhanu Yilma**

A Dissertation Submitted to the
Ethiopian Institute of Textile and Fashion Technology
in Partial Fulfillment of the Requirements for the Degree of
Doctor of Philosophy
in
**Textile Technology**

Under the Supervision of
Prof. Dr. Joern F. Luebben (MPE, Albstadt-Sigmaringen University, Germany)
Dr. Melkie Getnet (Ass. Prof.) (EiTEX, Bahir Dar University, Ethiopia)

Ethiopian Institute of Textile and Fashion Technology
Bahir Dar University

Bahir Dar, Ethiopia

August 2021

**Bibliographical information held by the German National Library**

The German National Library has listed this book in the Deutsche Nationalbibliografie (German national bibliography); detailed bibliographic information is available online at http://dnb.d-nb.de.

1st edition - Göttingen: Cuvillier, 2022



Nonnenstieg 8, 37075 Göttingen, Germany
Telephone: +49 (0)551-54724-0
Telefax: +49 (0)551-54724-21
www.cuvillier.de



1st edition, 2022

This publication is printed on acid-free paper.

ISBN 978-3-7369-7695-5
eISBN 978-3-7369-6695-6

## Declaration

I hereby declare that the dissertation submitted in fulfillment of the Ph.D. degree is my own work and that all contributions from any other persons or sources are properly and duly cited. I further declare that the material has not been submitted either in whole or in part, for a degree at this or any other university. In making this declaration, I understand and acknowledge any breaches in this declaration constitute academic misconduct, which may result in my expulsion from the programme and/or exclusion from the award of the degree.

Name: Baye Berhanu

Signature of Candidate:

Date: 08.09.2021

## Approval by Advisors

To the best of my knowledge and as understood by the student in the *Research Integrity and Copyright Disclaimer*, this dissertation adheres to the provisions of guidelines, policies and legislations of the Ethiopian Institute of Textile and Fashion Technology during his research work and use of copyrighted material. The dissertation is complete and can be presented to the dissertation evaluation committee.

Prof. Dr. Joern F. Luebben
(Principal advisor Name)

Signature

Dr. Melkie Getnet Tadesse
(Co-advisor Name)

Signature

## Approval by Examiner

I certify that I have supervised / read this study and that in my opinion it conforms to acceptable standards of scholarly presentation and is fully adequate, in quality and scope, as a dissertation for the fulfillment of the requirements for the degree of Doctor of Philosophy.

| | | |
|---|---|---|
| Dr. Melkie Getnet | Ass. Prof. | |
| Supervisor/ Co-supervisor | Academic status | Signature |
| Dr. Vishnu Dorugade | Prof. | |
| External Examiner 1 | Academic status | Signature |
| Dr. Lodrick Wangatia | Assoc. Prof. | |
| External Examiner 2 | Academic status | Signature |
| Dr. Abera Kechi | Assoc. Prof. | |
| Internal Examiner Chairman | Academic status | Signature |

(Examination Committee member)

This dissertation was submitted to the Ethiopian Institute of Textile and Fashion Technology, Bahir Dar University and is accepted as fulfillment of the requirements for the degree of Doctor of Philosophy.

Postgraduate Studies and Project Development Director

| | | |
|---|---|---|
| | Ass. Prof. | |
| Postgraduate Studies Director | Academic Status | Signature |
| Abera Kechi Kabish (PhD) Scientific Director | Assoc. Prof. | |
| Scientific Director | Academic status | Signature |

## Acknowledgements

Firstly, I would like to acknowledge the German Academic Exchange Service or Deutscher Akademischer Austauschdienst (DAAD) and Engineering Education Capacity Building Program (EECBP) of the MoSHE for the funding of this research through the Homegrown Ph.D. Scholarship Programme, 2017, and further for two months' financial support from the Albstadt-Sigmaringen university. I also acknowledge Dire Dawa University and Bahir Dar university [EiTEX] for paying study leave salary and research budgets, respectively.

My special thanks are sent to Prof. Dr. Joern F. Luebben, my principal supervisor, for his useful guidance and unlimited support throughout my study to conduct scientific research. He offered me the priceless opportunity to conduct my Ph.D. study in his department of Material and Process Engineering. Without his professional support, the completion of my study focused on a field completely new to me like plasma technology would have been impossible. I would like to thank my former co-supervisor Prof. Dr. Govindan Nalankilli of EiTEX, for his important role and the scientific advice for two years.

I would like to express my sincere gratitude to my new co-supervisor Dr. Melkie Getnet Tadesse, EiTEX, Bahir Dar University, for his kind guidance, constant caring about my research progress, critical and timely constructive comments on the Ph.D. dissertation.

I gratefully acknowledge Mr. Ziad Heilani, Mr. Paul-Gerhard Ringwald, Mr. Christian Lehr, and all members of Albstadt-Sigmaringen university for their technical support. I would also like to thank Prof. Dr. Joern F. Luebben's secretary Ms. Silke Deufel, for her cooperation in all aspects. Also, I would like to express my sincerest thanks to all the participants in this dissertation work.

This Ph.D. dissertation is dedicated to my parents to whom I owe my education and all accomplishments in my life. I would especially like to thank from my heart also my beloved wife and my children for their great support me to be successful.

## Abstract

Although blended polyester/cotton (P/C) blend fabrics have many applications in the apparel industry, the problem of less than optimal wear comfort due to the hydrophobic nature of polyester fibers persists. This problem lies in thermo-physiological and tactile/sensorial comfort properties. To overcome this problem, this work focuses on the surface modification of P/C blended fabrics by plasma treatment. Since plasma treatment offers numerous advantages, including technological, economic and ecological benefits, over wet-chemical finishing. Moreover, plasma only modifies the surface without changing the bulk properties of the substrate.

Therefore, low-pressure plasma was applied in this work to P/C blended fabric to modify its comfort properties. Different types of gases and hexamethyldisiloxane (HMDSO) monomer were selected to perform surface cleaning/etching, activation and polymerisation in plasma. In this work, 65% of polyester and 35% of cotton blend fabric was treated in a vacuum plasma chamber with air, Ar and $O_2$, and mixtures of $O_2$/HMDSO, $N_2$/HMDSO, $H_2O$/HMDSO and HMDSO monomer alone to obtain a more hydrophilic P/C fabric surface. The Taguchi method was used to design the experiment and analyze the largest influential variables and optimal parameter levels mainly for the $O_2$ plasma due to its wide application. The comfort properties of the blended fabric were investigated using an objective comfort evaluation method. In addition, various characterization techniques were used to analyze surface morphology and chemistry of the samples.

The results revealed that oxygen plasma treatment improved the thermal comfort properties of the blended fabric, except for the air permeability in some test series. Specifically, wicking of the fabric increased at least by 43.25% and 37.63% within 5 minutes in warp and weft directions, respectively, while the thermal resistance decreased by at least 20.16%. Similarly, the surface resistivity of the fabric reduced by 35.5%, while hand-feel value decreased by 4%. All these changes were due to surface modification by the plasmas used. In addition, the SEM images showed increased surface roughness due to the plasma treatment.

In general, the results obtained in this study demonstrate the feasibility and potential of plasma technology in modifying the surface properties of textile substrates. Plasma treatment affects the surface both physically and chemically without altering the bulk properties of the materials. By selecting appropriate gases and monomers, the desired changes in the textile surface can be achieved. However, upscaling and transferring a laboratory batch process to a commercial process can also present some technical challenges. In particular, the optimum parameters must be defined for each plasma process and reactor.

## Kurzfassung

Obwohl Mischgewebe aus Polyester/Baumwolle (P/C) in der Bekleidungsindustrie viele Anwendungen haben, besteht das Problem des nicht optimalen Tragekomforts aufgrund der hydrophoben Beschaffenheit von Polyesterfasern weiterhin. Dieses Problem liegt in den thermophysiologischen und taktilen/sensorischen Komforteigenschaften. Um dieses Problem zu überwinden, konzentriert sich diese Arbeit auf die Oberflächenmodifikation von P/C-Mischgeweben mittels Plasmabehandlung, da die Plasmabehandlung zahlreiche Vorteile, einschließlich technologischer, wirtschaftlicher und ökologischer, gegenüber der nasschemischen Veredelung bietet. Zudem wird nur die Oberfläche modifiziert, ohne die Bulkeigenschaften des Substrats zu verändern.

Daher wurde in dieser Arbeit Niederdruckplasma auf P/C-Mischgewebe angewendet, um dessen Komforteigenschaften zu verändern. Verschiedene Gasarten und Hexamethyldisiloxan (HMDSO)-Monomer wurden ausgewählt, um die Oberflächenreinigung/-ätzung, Aktivierung und Polymerisation im Plasma durchzuführen. In dieser Arbeit wurde ein aus 65% Polyester- und 35% Baumwolle bestehendes Baumwollmischgewebe in einer Vakuum-Plasmakammer mit Luft, Ar und $O_2$ sowie mit Mischungen aus $O_2$/HMDSO, $N_2$/HMDSO, $H_2O$/HMDSO und HMDSO-Monomer allein behandelt, um eine hydrophilere P/C-Gewebeoberfläche zu erhalten. Die Taguchi-Methode wurde eingesetzt, um das Experiment zu planen und die größten Einflussvariablen sowie die optimalen Parameterniveaus hauptsächlich für das $O_2$-Plasma aufgrund seiner breiten Anwendung zu analysieren. Die Komforteigenschaften des Mischgewebes wurden mit einer objektiven Komfortbewertungsmethode untersucht. Darüber hinaus wurden verschiedene Charakterisierungstechniken eingesetzt, um die Oberflächenmorphologie und -chemie der Proben zu analysieren.

Die Ergebnisse zeigten, dass die Sauerstoff-Plasma-Behandlung die thermischen Komforteigenschaften des Mischgewebes verbesserte, mit Ausnahme der Luftdurchlässigkeit bei einigen Versuchsreihen. Insbesondere erhöhte sich die Dochtwirkung des Gewebes innerhalb von 5 Minuten in Kett- und Schussrichtung um mindestens 43,25% bzw. 37,63%, während der Wärmewiderstand um mindestens 20,16% sank. In ähnlicher Weise verringerte sich der elektrische Oberflächenwiderstand des Gewebes um 35,5%, während der Wert für das Handgefühl um 4% abnahm. All diese Veränderungen sind auf die Oberflächenmodifizierung durch die verwendeten Plasmen zurückzuführen. Darüber hinaus zeigten die REM-Bilder eine erhöhte Oberflächenrauigkeit aufgrund der Plasmabehandlung.

Im Allgemeinen zeigen die in dieser Studie erzielten Ergebnisse die Einsatzmöglichkeiten und das Potenzial der Plasmatechnologie bei der Veränderung der Oberflächeneigenschaften von Textilsubstraten. Die Plasmabehandlung wirkt sich sowohl physikalisch als auch chemisch auf die Oberfläche aus, ohne die Bulkeigenschaften der Materialien zu verändern. Durch Auswahl geeigneter Gase und Monomere lassen sich die gewünschten Veränderungen der Textiloberfläche erzielen. Die Hochskalierung und Übertragung eines Labor-Batch-Prozesses in ein kommerzielles Verfahren kann jedoch auch einige technische Herausforderungen mit sich bringen. Insbesondere müssen für jeden Plasmaprozess und jeden Reaktor die optimalen Parameter festgelegt werden.

## Table of Contents

## List of Tables

## List of Figures

## Acronyms

| | |
|---|---|
| AATCC | American association of textile chemists and colorists |
| AC | Alternating current |
| AFM | Atomic force microscopy |
| APP | Atmospheric pressure plasma |
| APPJ | Atmospheric pressure plasma jet |
| ASTM | American society for testing and materials |
| ATR-FTIR | Attenuated total reflection-Fourier transform infrared spectroscopy |
| BS EN ISO | British standard / European standard / international organization for standardization |
| DBD | Dielectric barrier discharge |
| DC | Direct current |
| DIN | Deutsches institut für normung |
| DTGS | Deuterated triglycine sulfate |
| EDX | Energy dispersive x-ray analysis |
| FAMOUS | Fabric automated modular and optimization Universal System |
| FAST | Fabric assurance by simple testing |
| FSTT | Fabric smart tactile tester |
| HEMA | 2-Hydroxyethyl methacrylate |
| HF | Hand-feel |
| HMDSO | Hexamethyldisiloxane |
| KES-F | Kawabata evaluation systems for fabrics |
| LF | Low frequency |
| LPP | Low pressure plasma |
| MFC | Mass flow controller |
| MMT | Moisture management tester |

| | |
|---|---|
| MW | Microwave |
| P/C | Polyester/cotton |
| PECVD | Plasma enhanced chemical vapor deposition |
| PET | Polyethylene terephthalate |
| RF | Radio frequency |
| RH | Relative humidity |
| S/N | Signal to noise ratio |
| Sccm | Standard cubic centimeter per minute |
| SEM | Scanning electron microscopy |
| TO | Tera-Ohmmeter |
| TSA | Textile softness analyzer |
| UV | Ultraviolet |
| XPS | X-ray photoelectron spectrometry |
| XRD | X-ray diffraction |

## Symbols

| | |
|---|---|
| Å | Angstrom |
| $A_{IR}$ | Absorbance |
| C | Heat loss by convection |
| $C_r$ | Value of the reference fabric for water vapor resistance |
| °C | Degree Celsius |
| $C_k$ | Heat loss by conduction |
| $C_{res}$ | sensible heat loss due to respiration |
| D | Stiffness |
| $d_p$ | Penetration depth |
| $E_{res}$ | Evaporative heat loss due to respiration |
| $E_{sk}$ | Heat loss by evaporation from the skin |
| hv | Energy of photon |
| K | Sensitivity constant |
| $K_B$ | Boltzmann's constant |
| M | metabolic rate |
| m | Mass |
| $m_e$ | Mass of electrons |
| $m_i$ | Mass of ions |
| n | Number of molecules of gas |
| Ne | New English Count |
| $n_i$ | Number of ions |
| $n_n$ | Number of neutral particles |
| P | Pressure |
| $P_{wo}$ | Actual water vapor partial pressure |
| $P_{wsat}$ | Water vapor saturation partial pressure |
| $P_{wv}$ | Relative water vapor permeability |
| R | Heat loss by radiation |

| | |
|---|---|
| $R$ | Universal gas constant |
| $R_{et}$ | Water vapor resistance |
| $R_t$ | Thermal resistance |
| $T$ | Absolute temperature |
| $T_e$ | Temperature of electron |
| $t_H$ | Temperature of measuring head |
| $T_i$ | Temperature of ion |
| $T_{IR}$ | Transmittance |
| $t_o$ | Temperature of the laboratory air |
| TS7 | Softness value |
| TS750 | Smoothness value |
| $u_o$ | Without a fabric sample |
| $u_s$ | With a fabric sample |
| $V$ | Volume |
| $V_{th,e}$ | Thermal velocity of electron |
| $V_{th,i}$ | Thermal velocity of ion |
| $W$ | External work |
| $X_{iz}$ | Degree of ionization |
| $\delta$ | Duty cycle |
| $\theta$ | Angle |
| $\rho$ | Gas density |
| $\rho_s$ | Surface resistivity |
| $\rho_v$ | Volume resistivity |
| $\varphi$ | Relative humidity |
| $\dot{m}$ | Mass flow |
| $\dot{V}$ | Volumetric flow |
| $\dot{V}_S$ | Volumetric flow at standard conditions |

# CHAPTER ONE

# INTRODUCTION

## 1.1 Motivation

Production and consumption of synthetic fibers such as polyester, nylon, acrylic, and polyolefin highly increase worldwide. In 2017, the consumption of synthetic fibers possessed the highest share of 64.2% compared to regenerated and other types of natural fibers as presented in Figure 1.1. These fibers can be used for clothing, construction, medical, home furnishing, automotive, filtration and other applications depending on the interest of end-users [1]. Among these synthetic fibers, polyester (PET) is the most popular which accounts almost 80% of the total synthetic fiber output and is widely used for clothing fabric due to its strength, lightweight, resistant to many chemicals, and resistance to shrinking and stretching [2, 3]. However, because of its molecular structure and chemical composition, PET fiber lacks many comfort related properties compared with natural fibers [4]. On the contrary, natural fibers have inherent characteristics that fulfil the comfort properties. In this regard, cotton is a typical natural fiber and it has good comfort properties. In fact, there is no perfect fiber having all desirable textile properties. Therefore, blending of PET and cotton fibers can bring an alternative technique to enhance the required qualities and reduce the undesirable features of the fibers [5].

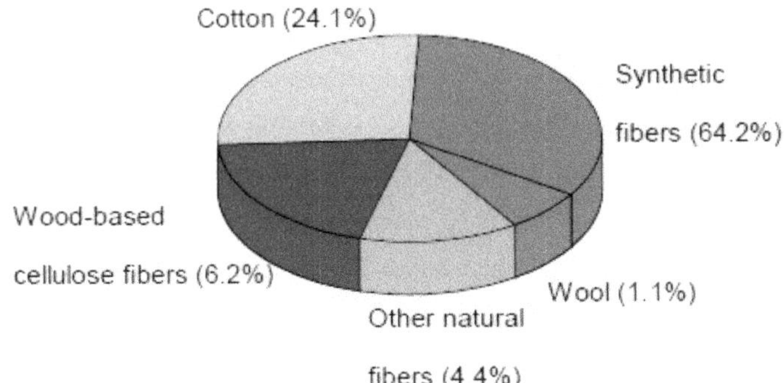


**Figure 1.1.** Consumption of fibers by the global market in 2017, adapted from [6].

In the textile industries, the polyester/cotton (P/C) blend fabric covers 58.45% of the world's market share [7]. This blending technology brought more attention in home furnishing and apparel applications. Nowadays, the demand of P/C blend fabric in the apparel industry has increased significantly due to the desired properties of user friendly, functional performance, aesthetic value, printability and affordability, which are more difficult to obtain altogether in both PET or cotton fabric products, separately. Despite the outstanding P/C fabric properties, PET surface hydrophobic character hinders the blend fabric comfort properties when functioned as an apparel fabric [8]. In addition, the comfortability of the P/C blend fabric mainly depends on the blend ratio and the environmental conditions [9]. Especially, when the proportion of the PET fiber is higher than cotton fiber the produced garment is uncomfortable for the wearers. Moreover, this problem would be more pronounced during outdoor wear.

As is already known, the hydrophobic nature of PET fiber has the potential to develop electrical resistivity in the fabric due to the tendency of the synthetic PET fiber to static electrification build up [10]. The built-up static charge, under low humidity conditions, does not dissipate for a long time because of the low conductivity of textile fibers. Consequently, accumulated static charges adversely affect the comfort properties of the garments when it rubs against the wearer's body, and at the same time, dust particles have the possibility to be attracted towards the fabric surfaces. The relationship between the fabric resistivity and its potential to build-up electrostatic charge is not always a clear-cut point; nevertheless, it is widely agreed that the most static-electrical problems emerging from the use of textile substrate can be reduced to manageable levels if the surface resistivity ($\rho_s$) and the volume resistivity ($\rho_v$) are dropped to $10^{11}$–$10^{12}$ Ω/square and $10^{7}$–$10^{8}$ Ω.m, respectively [11]. However, the PET fibers have insulating properties and have a volume resistivity of $10^{14}$ Ω.m [12].

Consequently, the hydrophobic characteristic of PET fiber affects the thermo-physiological comfort by disturbing the transmission of heat, movement of moisture vapor and perspiration from the skin to the surrounding environment through the blend fabric [13, 14]. This indicates that the types of blend component and thermal insulation property of the fabric as well as the abilities of its moisture management influence the clothing comfort. Indeed, in warm environments, the volume of sweat is considerably higher in hydrophobic fabrics when compared with hydrophilic fabrics [15]. Particularly, when the PET fiber content increases in the blend, the water vapor permeability of the fabric decreases linearly. This is due to the decreasing absorption of the water molecules in PET fiber. Under this situation, the body is unable to maintain a steady-state of body temperature within its own heat transfer mechanism, and then, it starts to sweat. Hence, evaporation became low from the fabric. Subsequently, the water vapor is condensed in-between the fabric and skin ahead of flow down on the skin in liquid form. This liquid would wet the fabric and clung with the body. Finally, the wearer feels discomfort [4]. To prevent this and above-mentioned limitations, the P/C blend fabric surface must be treated to give better clothing comfort.

Therefore, the surface treatment of P/C blend fabric is necessary to improve the comfort properties, while retaining its excellent blend fabric properties. In this regard, several techniques have been applied to fabric surface modification, focusing on the enhancement of its surface to be more hydrophilic. These include wet chemical processing, plasma treatment, corona discharge and flame treatment [16]. However, the above mentioned traditional chemical-based treatments have technological and ecological drawbacks such as changing the bulk properties, bringing toxicity to the environment and are not cost effective [17]. In addition, corona and flame treatment can bring contamination and surface non-specific problems, respectively. On the other hand, to improve the fabric static charge dissipation or to eliminate the generations of static electricity, different treatment methods have been performed. These methods have been used to introduce the conductive metal/organic fibers into woven structures [18], radioactive static eliminators [11], and antistatic agents [19]. Nevertheless, introducing the conductive metal fibers into the woven structure also affects the mechanical properties of the fabric. To resolve these problems, plasma treatment is an ideal option.

Plasma technology provides an attractive approach in surface treatment of textile materials as it is considered as a versatile, dry, eco-friendly and resource-saving process as it uses null or minimum amount of water and chemicals [20]. So far, it merely modifies the outermost surface of the textile substrate without changing its bulk properties [21]. In particular, low-pressure plasma system is usually between 0.01 and 10 mbar, which is more preferred due to the high concentration of reactive species, the uniformity of a large surface area, high controllability, and the superior chemical selectivity and reproducibility of the results [22, 23]. This technology has been found to be effective to impart functional properties of several textile fibers, but comfort properties of the P/C blend fabric are not studied yet. Therefore, in this research, the low pressure plasma was implemented to provide a more hydrophilic surface of P/C blend fabric for shirt clothing application. In this study, the researcher selected the blend of 35% cotton and 65% polyester because it is the most popular blend fabric and has been used for many applications [24], such as sportswear [25] and medical protective clothing after finishing treatment [26].

## 1.2 Objectives of the research

The general objective of this research is to study the influence of plasma surface modification on comfort properties of P/C blend fabric by considering the climatic conditions. Particularly, the intention of this study is to improve the hydrophilicity of shirt cloth, made from 65% of PET and 35% cotton, by modifying the hydrophobic characteristic of its surface with low-pressure plasma. Even though the hydrophobic nature mainly belongs to PET fibers, the plasma treatment can change both the surface of PET and cotton fibers. The reasons for using shirt cloth are due to its wide application for uniform clothing and used next to underwear. In addition, it has a great chance directly to contact the nude body around the neck, shoulder and arms. The specific objectives of the research are:

- To utilize gases and monomer suitable for plasma surface modification medium;
- To evaluate thermal and tactile comfort properties of P/C blend fabric including air permeability, wickability, water absorbency, electro-physical, hand-feel and other properties;
- To investigate the aging behavior of plasma-treated P/C blend fabric;
- To characterize surface morphology and surface chemistry of P/C blend fabric;
- To identify undesirably affected comfort properties of plasma-treated fabric, experimentally.

## 1.3 Structure of dissertation

**The structure of the dissertation approach is organized as follows:**
This dissertation focuses on the surface modification of P/C blend fabric by applying low-pressure plasma treatment and consequently studying the comfort characteristics of the treated sample.

**Chapter 2** illustrates the basics about clothing comfort and plasma surface modification of textile substrates. First, a brief introduction to clothing comfort is given and key concepts regarding thermo-physiological, physical, and aesthetic comfort as well as assessment of comfort properties are elaborated. Secondly, introduction to fundamentals of plasma technology, mechanism of plasma-textile surface interaction and application of plasma in textiles using gases and liquid monomers are explained.

**Chapter 3** describes the set-up of the vacuum reactor for plasma surface modification of the P/C blend fabric which was used in this research work. In addition, the fabric specification of the blend fabric (shirt cloth) and chemicals, consumed for plasma treatment, are mentioned. Moreover, the experimental procedures of the comfort assessment methods and surface characterization techniques were explained.

**Chapter 4** discusses the experimental results of plasma surface modification of the P/C blend fabric. The effect of oxygen plasma on thermal comfort properties of the fabric was studied using Taguchi methods. The influence of oxygen, argon and air plasmas on hand-feel and electro-physical properties of the fabric were also evaluated. On the other hand, the effect of the mixture of reactive gases/molecules and Hexamethyldisiloxane (HMDSO) plasmas on tensile strength and pilling resistance properties of P/C blend fabrics were discussed. In addition, the effect of aging on oxygen, argon and air plasmas treated blend fabric was investigated. The surface morphology and surface chemistry of plasma treated and untreated P/C blend fabric was characterized by using scanning electron microscopy (SEM), attenuated total reflection Fourier transform infrared spectroscopy (ATR-FTIR), and energy dispersive x-ray spectroscopy (EDX).

**Chapter 5** presents the conclusions and future works of the research. The conclusion summarizes the findings and contribution of this study. It is drawn from the results of the experimental investigations. Additionally, the future work indicates the outlook that will be continued using the results of this dissertation as a background information.

The overall structure of the dissertation can be outlined in Figure 1.2.

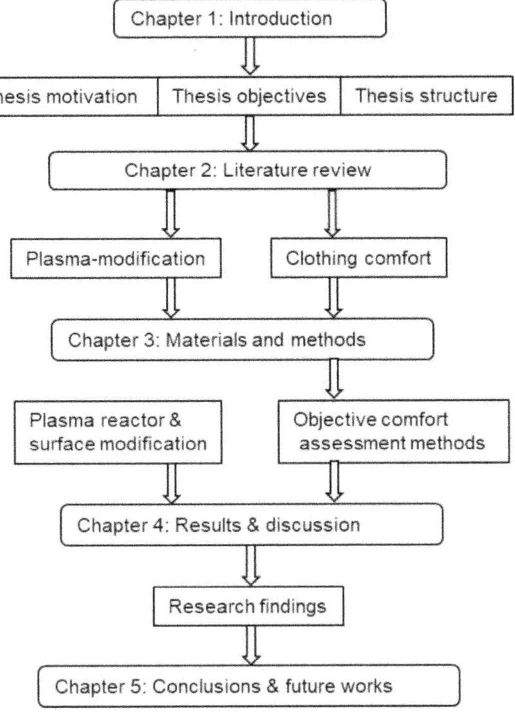


**Figure 1.2.** Dissertation structure.

# CHAPTER TWO

# LITERATURE REVIEW

In this chapter, the fundamentals of plasma process and clothing comfort are illustrated and discussed. Here, in addition to the scientific and practical information, other related fundamental evidence is given to realize the concept clearly. However, more attention is given to plasma modification and clothing comfort.

## 2.1 Fundamentals of plasma process and its surface modification

### 2.1.1 Introduction to plasma

Plasma is a partially ionized form of gas and often also referred to as the fourth state of matter, next to solid, liquid and gaseous state. In phase transitions, when a solid is heated enough, it can melt into liquid or sublime into gases. After obtaining sufficient energy, the particles in a liquid escape from it and vaporize to gas. Subsequently, when a significant amount of energy is applied to the gas, atoms and/or molecules collide with each other and knock their electrons off in the ionization process. Eventually, the higher number of electrons and ions can change the electrical property of the gas, which consequently becomes ionized gas or plasma [27]. This phase change of states of matter is illustrated in Figure 2.1. Atoms and molecules can be ionized or charged either thermally, magnetically or electrically [22, 28]. The plasma state comprises approximately an equal number of positively and negatively charged particles, i.e. quasi-neutrality, radicals, excited states, metastables, atoms, molecules, electrons, and photons. This state of matter was first noted by Sir William Crockes in 1879, and Irving Langmuir named 'plasma' in the 1920s [29]. Plasma differs in various respects including density of charged particle, pressure, temperature, and the presence of external electric and/or magnetic fields.

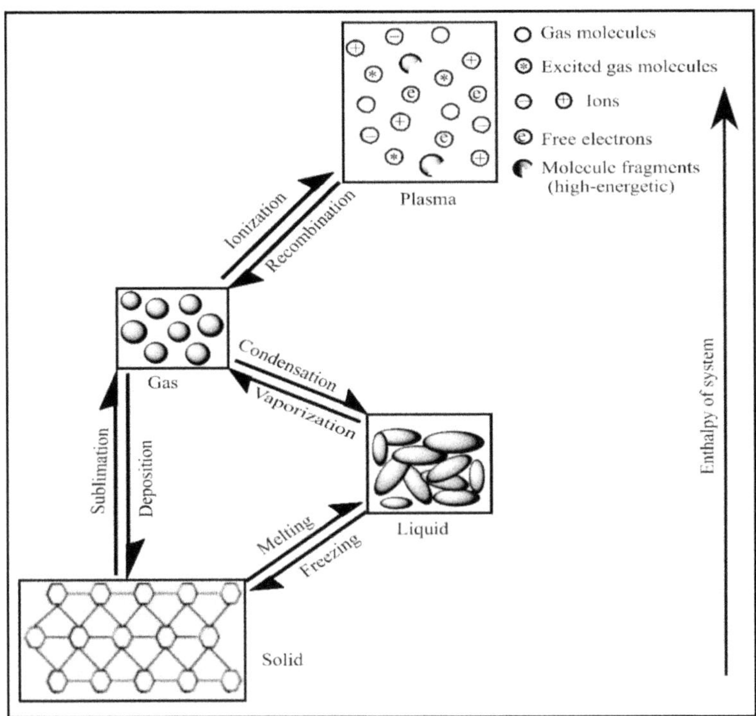


**Figure 2.1.** Phase changes of state of matter, adapted from [27].

Based on the temperature of electrons, ions and neutrals, plasma can be classified into thermal (hot) and non-thermal (low-temperature or cold) plasma, as shown in Figure 2.2. Thermal plasmas are associated with electric arcs, thermonuclear reactions, and laser-induced reactions [30]. In thermal plasma, all the species (i.e. electrons, ions and neutrals are at a similar temperature) that composing the plasma are in the thermodynamic equilibrium. The temperature of the plasma species produces a high volume of heat (>1,500 ℃) and is used as a source thereof. It is only used for changing the bulk material properties, being not suitable for textile and nearly all other materials. Cold plasma, on the other hand, is maintained at around ambient temperature, or somewhat greater than it (20-250 ℃). This is due to the electron temperature being much higher than ion and neutral temperature. It can be successfully applied to textile processing because textile materials are generally more heat sensitive polymers than metals and ceramics [16, 30, 31].

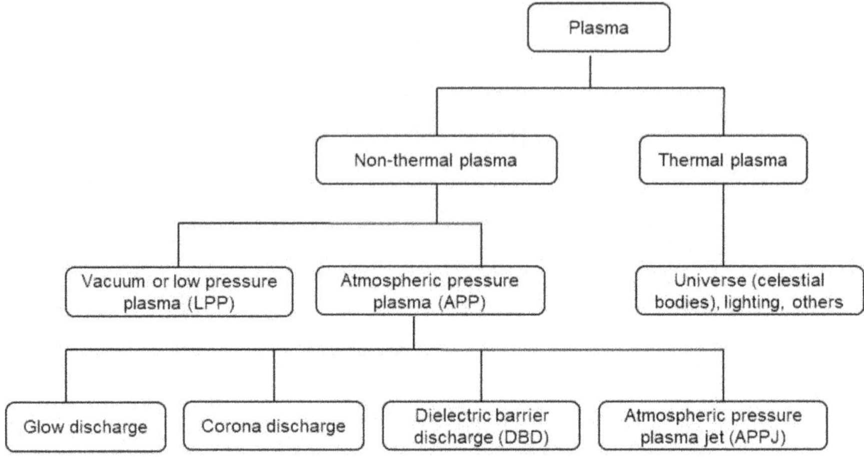


**Figure 2.2.** Classification of plasma, adapted from [31].

Cold plasma can be done in a vacuum under low pressure or at atmospheric pressure, both of which are used for textile surface modification. As mentioned above, cold plasma contains many reactive species (i.e. electrons, ions and neutrals), and these can initiate physical and chemical reactions on the surface of textile substrates [21]. Such modifications are limited to a few nanometers in depth, which only changes the outermost chemical, structural and surface properties of a substrate. Thus, plasma treatment can impart a desired property to textile fibers. The effectiveness of surface modification depends on plasma conditions and types of treated materials. Plasma treatment is very suitable, versatile, and multifunctional in textile processes [28, 32].

### 2.1.2 Generation and properties of plasma

**Generation of plasma**

The plasma state is generated by supplying energy to a neutral gas and creating the formation of charge carriers [33]. When energetic electrons and/or photons collide with the neutral molecules and atoms in the feed gas (electron-impact ionization or photoionization), a number of electrons and ions can be produced in the plasma zone as illustrated in Figure 2.3. The plasma can be created in laboratory and industry scale by increasing the amount of energy to a neutral gas irrespective of types of energy source. In particular, it can be generated by electrical (electric discharges), mechanical (adiabatic compression, shock waves), thermal (electrical heated furnace, intense redox reactions like flames), chemical (exothermic reactions), radiant (high energy electromagnetic radiation, particle radiation) and nuclear (fission, fusion) energies as well as by the combination of those mechanisms [33, 34]. All these energies can generate plasmas for processing purposes by different principles according to their energy sources. In addition to energy, at various pressures, including low or vacuum, atmospheric, and high pressure, plasma generation can be achieved [35].

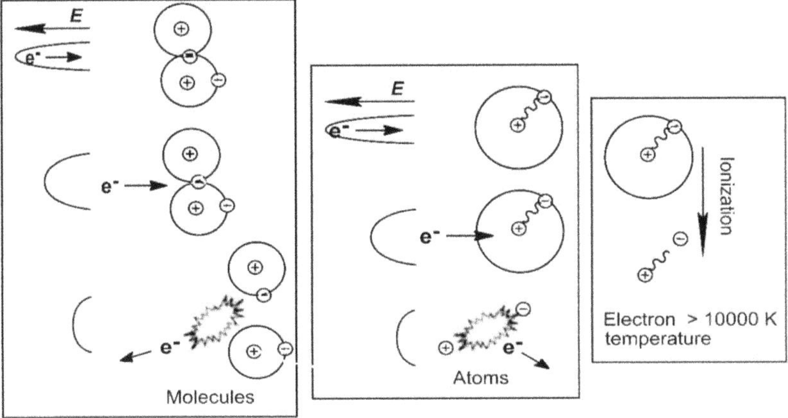


**Figure 2.3.** Generation of plasma by energy transfer (E) and accelerated **e**⁻, adapted from [36].

Among aforementioned types of energy, the electrical field is the most suitable and widely used method to generate and sustain the plasma state. Direct current (DC) and alternative current (AC) power supplies have been used for the technological and technical application by generating cold plasma at lower temperature [37]. However, AC is more appropriate for plasma generation. Because it can be operated in a wide range of frequencies from radiofrequency, RF (kHz) to a microwave, MW (MHz and GHz) ranges. Specifically, there are three primary frequencies used in plasma surface treatment. These are 40 kHz, 13.56 MHz and 2.45 GHz. Typically, the frequencies of 13.56 MHz and 2.45 GHz in the MW range are most commonly used for plasma processes, in research and industry, to avoid interference with telecommunication application [38]. However, a low frequency, LF (40 kHz) can make a great difference in quality. At this frequency, the highest ion density, i.e. more number of plasma particles per square inch, can be achieved compared to the other two frequencies [39]. In this work, LF discharge is applied and shall be discussed more detail in chapter three.

**Plasma ionization**

Ionization is the first and the key process in plasma. The degree of ionization ($X_{iz}$) of a plasma is a measure of the number of charged particles, i.e. ionized atoms and molecules, to the total number of particles including neutrals ($n_n$) and ions ($n_i$) [40], and it is defined as:

$$X_{iz} = \frac{n_i}{n_i + n_n} \tag{2.1}$$

Only a small fraction of atoms and molecules are ionized in an electric discharge of cold plasma. Typical 1 in $10^5 - 10^6$, at a gas pressure of 133 $p_a$ corresponds to a particle density of $10^{22}$ $m^{-3}$ and an electron density of $10^{16}$ $m^{-3}$. For technical plasmas, the $X_{iz}$ is typically in the range of $10^{-6}$ to $10^{-3}$.

**Plasma and sheath formation**

The plasma sheath exists between the actual plasma and internal wall of a reactor. In other words, the plasmas, at quasi-neutral (approximately equal number of positive and negative charges), create positively charged boundary layers when close to a material or confining wall surfaces. It is a so-called sheath with a positive space charge. This is correlated to the higher thermal velocity of electrons ($V_{th,e} = (2K_BT_e/m_e)^{1/2}$), $K_B$ is Boltzmann's constant, compared to the velocity of ions ($V_{th,i} = (2K_BT_i/m_i)^{1/2}$), in the cold plasma, by two orders of magnitude because of the imbalance of their mass ($m_e << m_i$) [34]. The positively charged sheath is the results of potential gradient profile as exemplified in Figure 2.4.

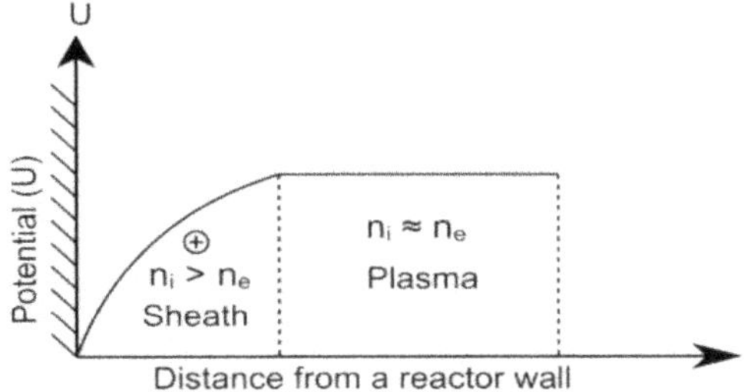


**Figure 2.4.** Illustration of electric potential and sheath formation near a reactor wall, adapted from [40].

Electrons promptly drift to an absorbing wall in a short period of time after plasma bombardment due to their high mobility. Consequently, the ion sheath develops a potential gradient towards the walls of a reactor. The resulting potential is directed to the walls. As a result, more electrons are repelled and confined in the plasma zone, whereas the positive ions penetrating the plasma sheath are raced up by the electric potential (U) to the walls [40, 41].

**Collision reactions**

Ionization and excitation processes are necessary to generate and sustain a plasma via momentum, charge and energy transfer in elastic or non-elastic collision reactions. In elastic collision, the kinetic energy is moved into the collide partners, but the total of kinetic energy is conserved and internal energy is not affected. On the other hand, the kinetic energy, in non-elastic collisions, is transferred into potential energy. Consequently, the sum of kinetic energy is not conserved. Most commonly, dissociation, ionization and excitation processes are the results of inelastic collisions. When the excitation energy of a metastable atom surpasses the ionization internal energy of another atom, their collision leads to penning ionization. Table 2.1 reviews the most important collision reactions including electrons, ions, neutrals and heavy species that can occur in plasma processes. In cold or weakly ionized plasma, the collisions between electrons and neutral species are frequently dominant. Moreover, particles excited by collisions with charged species can go through radiative de-excitation and recombine processes [36, 40].

**Table 2.1:** Collision reactions among electrons, ions, neutrals and heavy species.

| *Description* | *Reactions* |
| ---: | --- |
| Elastic scattering | $e^- + A \longrightarrow e^- + A$ |
| | $B^+ + A \longrightarrow B^+ + A$ |
| Dissociation & fragmentation | $e^- + AB \longrightarrow e^- + A + B$ |
| | $A^+ + BC \longrightarrow A^+ + B + C$ |
| Dissociative ionization | $e^- + AB \longrightarrow 2e^- + A^+ + B$ |
| Dissociative attachment | $e^- + AB \longrightarrow 2e^- + A^- + B$ |
| Ionization | $e^- + A \longrightarrow 2e^- + A^+$ |
| Excitation | $e^- + A \longrightarrow e^- + A^*$ |
| Penning ionization | $e^- + A^* \longrightarrow 2e^- + A^+$ |
| Charge exchange | $B^+ + A \longrightarrow B + A^+$ |
| Chemical reaction | $A + BC \longrightarrow C + AB$ |
| Recombination | $e- + A^+ + B \longrightarrow A + B$ |
| De-excitation processes | $e^- + A^+ \longrightarrow e^- + A^* + hv$ (energy of photons) |
| | $A^* \longrightarrow A + hv$ |
| | $A + BC \longrightarrow ABC + hv$ |
| | $A^* + hv \longrightarrow A + hv + hv$ |

### 2.1.3 Parameters of plasma treatment

Plasma and process parameters are crucial in customizing the surface properties of polymers and textile materials. Indeed, plasma parameters such as, plasma density, temperature and energy distribution of electron and ion depend on the process parameters which include types of plasma forming gas, process pressure, exposure time, gas flow rate, discharge power and others [42, 43]. The effectiveness of plasma treatment depends on the selection of optimal plasma conditions include the type of gas/monomer, flow rate, pressure, power, treatment time and types of materials to be treated [28].

The results of plasma treatment strongly depend on the types of gas used in the plasma process. Each gas produces a unique plasma composition and surface properties. For instance, oxygen, carbon dioxide, ammonia and nitrogen plasmas impart hydroxyl, carboxyl and amine groups, respectively, to the polymer surface and enhance hydrophilic property [44]. Plasma intensity is also determined by a combination of operating pressure and discharge power. When the input power increases under a constant pressure, the total amount of excited particles and their energy level increased inside the plasma accordingly. In plasma process, the exposure time plays an important role. The polymer and textile surface will be negatively affected by the prolonged treatment time. Therefore, careful control of treatment time is required [43]. Similarly, textile materials used for plasma treatment has their own effects. For example, for the porous textile, active plasma species can pass through a number of pores and modify the inner surfaces as well [45].

### 2.1.4 Plasma surface modification and characterization

Textile surface modification can be done by conventional and/or advanced innovative methods. In this regard, several techniques have been applied on the fabric to impart the desired properties during the finishing process. However, in wet chemical surface modification, the textile surface is mostly treated in aqueous solutions. In this process, a high amount of water, chemical and energy consumptions, and long treatment time takes place. Moreover, recently, environmental regulation is tightened regarding aqueous based effluents from the textile industry [46]. In flame, treatment has been performed at elevated temperature (1000 – 2000 $^{O}$C), but the excited species form surface heterogeneity and reduces optical clarity of the polymers [47]. In addition, Corona discharge treatment is limited to a small area, and the UV radiation may affect the optical property of polymer and UV light transfer. This effect leads to inconsistency of treatment [44].

In order to overcome these detrimental effects on the textile substrate and environment, the textile industry is highly motivated to find an alternative dry finishing process that is an ecological and economical method. Plasma surface modification may accomplish the needs of the textile industry as it is considered as a green and dry process. Nowadays, plasma surface modification has been given more attention in the textile fields, as it looks to be an alternative to a classical textile finishing process. Plasma surface treatment is an ergonomically simple process, which is solvent-free, clean, cost-saving, safe and eco-friendly.

Moreover, through plasma treatment it is possible to impart a typical textile finish without altering the bulk properties of the textile material, which is difficult to achieve while using other techniques. For example, Persin et al. [48] studied the sorption characteristics and surface properties of modal, viscose and lyocell regenerated cellulosic fabrics by using conventional wet-chemical pretreatment and low-pressure-oxygen plasma treatment, separately. In fact, both treatments modified the surface chemistry of cellulose fabrics. However, the content of carboxylic acid was increased ~ 4.8% by the wet-chemical pre-treatment, whereas plasma treatment increased it by ~ 9.7%. The authors concluded that oxygen plasma treatment has a significant effect on the surface energy as well as on the polarity of cellulosic fabrics. In another study, Demir [49] reported on the air/Ar atmospheric plasma treatment of mohair fibers to examine its hydrophilicity, fiber to fiber friction, grease content, dyeing, shrinkage and color fastness properties. The results indicated that the hydrophilicity, fiber friction coefficient, dyeability and shrinkage properties of the mohair fibers were enhanced after plasma treatment. In principle, all kinds of fibers and polymers can be treated by plasma [22].

In cold or low temperature plasma, the high-energy electrons and low-energy molecular species interact with the surface of the substrate. Due to the presence of charged particles, pressure, temperature and electric and/or magnetic fields, disruption of the chemical bonds occurs in the surface of treated materials. Consequently, surface morphology and surface chemistry of a fiber or fabric is altered. This includes, changes in physical and chemical properties of surface layers/structures either temporarily or permanently. Actually, plasma modification is restricted to the uppermost, 1–100 nm, surface of the substrate to be treated and plasma conditions [50, 51]. Here, different types of analytical tools can be used to characterize plasma modified surfaces depending on the nature of modification and the resources available. The common analytical tools, used for physical and chemical characterization of surface modification, can be grouped into microscopic, spectroscopic and non-spectral techniques. Under microscopic techniques, SEM and atomic force microscopy (AFM) are typical examples, whereas in spectroscopic, X-ray photoelectron spectroscopy (XPS) and ATR-FTIR are well-known methods. In addition, water contact angle, liquid absorptive capacity and capillary rise methods belong to non-spectral techniques [44, 52]. Surface characterization techniques which were used in this work will be elaborated in chapter three.

### 2.1.5 Mechanism of plasma-textile surface interaction and its effect

Before discussing the effect of plasma on textile substrates, it is very important to understand the interaction mechanisms between the plasma species and textile surfaces. In the bulk plasma, reactive species, such as radicals, ions, electrons, excited particles and photons are generated by means of fragmentation, dissociation, ionization, and excitation through electron collisions and photochemical processes. These species led to physical (chain scission) and chemical (cross-linking) interactions with the textile surface depending on plasma parameters, namely types of gas/monomers used, discharge power, flow rate, system pressure, exposure time and others [53]. Figure 2.5 illustrates the interaction of plasma species with textile surfaces under vacuum plasma. The produced reactive particles react directly with the surface by retaining the bulk properties of the treated materials. Characteristically, these interactions result in physical (etching, surface morphology) and chemical (grafting, polymerization) modifications on the substrate surfaces. In fact, the surface modification is confined up to 1000 Ångström (Å) [35].

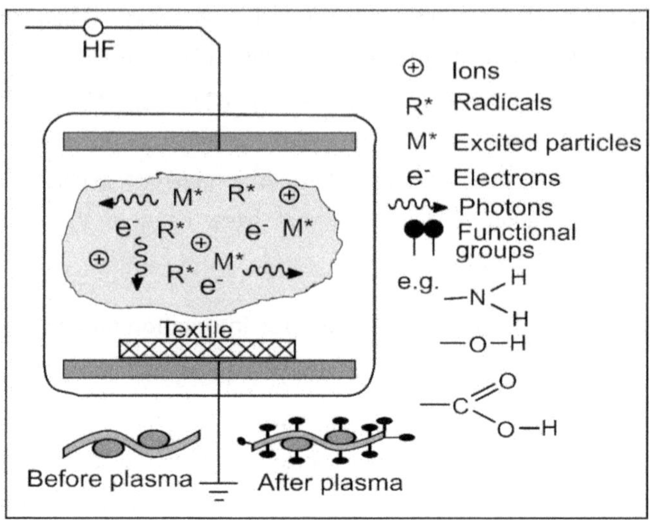


**Figure 2.5.** Mechanisms of plasma-textile surface interactions, adapted from [48].

Essentially, after plasma-textile surface interactions, four major effects can be observed on the surfaces depending on the plasma conditions and types of material to be treated. However, more than one effect is usually present to some extent, but one may be dominated over the others. These effects are surface cleaning or etching, surface activation, grafting and polymerization [30].

**Surface cleaning and etching**

Plasma treatment is a surface phenomenon and modifies only the outermost layer of a substrate surface [22]. Impurities and contaminants, such as oils and greases are removed from a fiber surface by physical etching and chemical reactions in the form of volatile products, such as water vapor and carbon dioxide. Novel gases (argon, helium), nitrogen and oxygen plasmas are primarily used for etching of polymer surfaces. Argon is the most well-known inert gas used in plasma etching due to its high ablation efficiency, chemical inertness to the surface, and comparatively cheaper [54]. Oxygen-containing plasmas can be applied to remove organic contaminants, viz. oligomers, anti-block agents and antioxidants [55]. The extent of weight loss mainly depends on the nature of substrate and plasma energy. Particularly, polymers containing oxygen in the functional groups (esters, ethers, carboxylic acids, ketones) are the most sensitive, whereas polyolefins containing hydrocarbons are less sensitive [56]. The only difference between etching and cleaning is the amount of materials that are removed from the surface [55].

**Surface activation**

Plasma surface activation is usually performed by means of non-polymerising inert and reactive gases, that don't have carbon, at the surface of textile substrates. Inert gases like helium and argon primarily initiate surface activation by producing free radicals through chain scission, while reactive gases, such as oxygen, nitrogen and ammonia can introduce oxygen or nitrogen containing functional groups [57]. In oxygen plasma, two kinds of effects can occur simultaneously. The first one is etching of fiber/polymer surfaces by the reaction of atomic oxygen with the surface carbon atoms. The other one is the formation of oxygen functional groups at the surface by means of the reaction of plasma species with surface atoms [43]. Subsequently, the surfaces with low functionality become more reactive by increasing those polar groups at the surface. Consequently, the surface energy of a substrate is increased. It is a significant characteristic especially, for wettability of a material for various processes. In this work also argon, oxygen, nitrogen, air and water vapor were used as plasma gases.

**Plasma polymerization**

Plasma polymerisation is a process of thin film formation onto the surface of a given substrate by using organic monomers, such as methane, ethane, tetrafluoroethylene, 2-hydroxyethyl methacrylate (HEMA) and HMDSO [22, 56]. It is formed when a monomer is introduced into a plasma reactor and gains energy by means of inelastic collision. Then, the monomer is excited and fragmented into activated small molecules. Subsequently, these small molecules are de-excited and recombined to form a large molecule of thin layer [46]. However, the structure of a plasma–deposited film layer is highly complex and governed by many factors, including reactor design, power level, temperature of substrate, system pressure, flow rate and structures of monomers [53]. Nevertheless, plasma polymerisation gives a high degree of cross-linking and great adherence to the surface of the substrate compared to conventional polymerisation [22]. In this work, HMDSO monomer was used for plasma coating of P/C blend fabric. The main reason for selecting HMDSO is its stability, non-toxicity, low flammability, accessibility, and much safe to the environment [58, 59].

### 2.1.6 Application of plasma in textile

Recently, many books [16, 43, 60–62] and extensive review articles [22, 28, 35, 63, 64] have been published regarding application of plasma on different textile substrates. Plasma treatment is perhaps the most versatile treatment technique to develop the unique surface properties used for various applications. The most common applications method includes imparting hydrophilic or hydrophobic properties, surface cleaning, modification of surface morphology/topography, enhancing adhesion, improving printability and dyeability, and finishing [63].

The effect of plasma treatment has to optimise the interaction between the fiber surface and the finishing product that is added to the bath. The treatment effects are achieved by bringing about surface modifications on a micro- or nano-scale without changing the bulk properties of the textiles. The effect can be advantageous in two ways: (i) by improving wetting properties of the textile product, and (ii) by enhancing the interaction between the finishing product and fiber surface, i.e. less of the finishing product is needed. The properties of a plasma-treated and finished textile can also be enhanced, as compared to a textile merely finished by the classical method to which the same amount of the finishing product is added [65].

Plasma treatment with different kinds of plasma gases and/or monomers can impart unique functionalities to textile substrates. A lot of research has been done on this and reported in the literature. Shahidi and Ghoranneviss [66] reported that low-pressure oxygen plasma treatment completely sterilised cotton fabrics inoculated with various concentrations of staphylococcus aureus. This finding explained by the fact that highly energetic UV light and activated free radicals generated during plasma treatment weakened the cell wall of the microorganisms by reacting with the hydrocarbon bonds, causing the disruption of unsaturated bonds, particularly the purine and pyrimidine components of the nucleoproteins. The authors suggested that oxygen plasma can be effectively used as an alternative method for sterilising and protecting cotton fabrics. Vesna et al. [67] studied corona treatment for fiber surface activation that can facilitate the loading of Ag nanoparticles from colloids onto polyester and polyamide fabrics and thus enhance their antifungal activity against candida albicans. Polyester and polyamide fabrics pretreated by corona treatment loaded with Ag nanoparticles showed better antifungal properties compared to untreated fabrics. In addition to sterilisation, plasma treatments can also impart antimicrobial and antibacterial functionality or aid in antimicrobial finishing [68, 69].

Cold plasma treatment can be used as an effective method for modifying the surface properties of wool fiber by overcoming the drawbacks of the conventional wet-process. Plasma treatment could impart significant shrink-resistance and anti-felting properties to wool fabric. The shrink proofing of wool fabric by plasma, both at atmospheric pressure [70] and at vacuum pressure [71], was initiated to replace the classical wet-processes that cause various degrees of environmental stress.

The resistivity of textile material can be reduced by introducing functional groups at the fiber surface through plasma treatment and by forming hydrogen bonds with atmospheric water. Kan and Yuen [72] studied the relationship between the moisture content and half-life decay time for the static properties of PET fabric. The results showed that an increment in moisture content consequently resulted in shortening the time for the dissipation of static charges. Moreover, there was a great improvement in the anti-static property of oxygen plasma-treated polyester fabric after compared with that of polyester fabric treated with a commercial anti-static finishing agent. Rashidi et al. [73] investigated the effect of low pressure air plasma on the surface resistivity of cotton and PET fabrics. The surface resistivity of cotton and PET is dramatically reduced after plasma treatment. Similar research works related to surface resistivity have been published by different authors [17, 74].

In addition, there are a number of excellent review papers that included the impact of unique functionalities to textile materials such as wrinkle resistance, the self-cleaning function, UV protection, flame retardancy, etc. by applying plasma treatment in textile finishing [22, 28, 64]. Moreover, plasma treatment is capable of changing the physical properties of the fabric like air permeability, water vapor permeability, electro-physical property, fabric hand properties, thermal properties and others. These physical properties directly or indirectly correlated with clothing comfort. The influence of plasma on comfort properties of different textile materials is also discussed hereafter.

### 2.1.7 Effect of plasma treatment on comfort properties of fabrics

For apparel fabrics, defining of comfort in a single sentence is very difficult and complex because it has multi-disciplinary nature [75]. However, many researchers have tried to define it differently according to their understandings [76]. The most acceptable definition of comfort was given by Slater. He explained that comfort is the pleasant state of physical, psychological and physiological harmony between the environment and human body. All these aspects make equal contributions the wearer to feel comfort [77].

Recently, plasma treatment has been used to study the comfort properties of textile fibers/fabrics, including tactile and thermal properties, by means of surface modification using various gases and monomers. For instance, the study by Rombaldoni et al. [78] has depicted that HMDSO plasma polymerisation improved the thickness of a wool fabric due to the deposition of a thin polymeric film and induced roughness of the wool fabric surface. Consequently, these effects reduced the air permeability of wool fabric. In another study, Ferri et al. [79] reported that nitrogen atmospheric-pressure plasma treatment significantly affected the thermal properties of plain-wool fabric. After plasma treatment, the results showed that the thermal resistance and thermal diffusivity increased, whereas the thermal absorptivity and volumetric heat capacity decreased compared to untreated wool. The effect of plasma treatments on the low-stress mechanical properties of wool [80], polyamide [81] and polypropylene [82] fabrics have been studied. A similar investigation performed to find the significance of plasma treatment to zari silk yarn on the comfort properties of the fabric [83].

Prakash et al. [84] studied the influence of oxygen atmospheric-pressure plasma on water-vapor and air permeability of single jersey bamboo fabric. Their results revealed that water-vapor permeability increased, whereas the air permeability decreased. The improvement of water vapor transfer by plasma treatment would provide better comfort properties to the fabric. Similar outcomes were observed after treatment of bamboo fabric with air plasma [85]. The effect of oxygen, air and argon plasma treatment on the thermal comfort properties of woven-cotton fabric has been investigated. The air permeability of fabric becomes reduced along with plasma treatment and oxygen plasma has the highest influence on it. On the other hand, the thermal resistance and water vapor permeability of the fabric increased with the plasma treatment [86]. A comparable result also was observed with air plasma treated cotton fabric in another research work [87]. The plasma treated polyester fabrics revealed better fabric characteristics specifically in terms of wickability, water vapor permeability and antibacterial activity compared to untreated fabrics. The SEM micrographs also showed the modified surface of polyester fabric [88].

However, as the literature review shows only a few studies were carried out regarding the influence of plasma treatment on comfort properties of fabric. In addition, these research work focused on merely one type of fibers rather than blend fabrics. The comfort properties of plasma treated-polyester fabric are not fully investigated. Nevertheless, in this dissertation, a comprehensive study of the comfort properties of a P/C (65/35%) blend fabric treated in low-pressure plasma was carried out. In this study, the effect of plasma treatment on various comfort properties of P/C blend fabric, such as air permeability, water vapor permeability, wickability, fabric hand, thermal, electro-physical and other surface related properties, were investigated.

## 2.2 Clothing comfort

### 2.2.1 Aspects of comfort

Comfort is an essential and universal need of human beings [89]. Barker described comfort as being a function of clothing variables and physical properties of materials as well as it has to be understood within the whole situation of human psychological and physiological responses [90]. Nielsen also explained comfort in relation to physical sense as the human body being in thermal comfort zone by maintaining the heat balance with the surrounding environment. Additionally, the design of cloth must be ergonomically compatible without pressure and unpleasant contact with the body in order to prevent skin irritation in work clothing [91]. Moreover, Bartels elaborated the wear comfort by dividing into four main aspects. These are: (i) thermo-physiological, (ii) skin sensorial, (iii) psychological, and (iv) ergonomic wear comforts [92]. A similar classification of comfort is done by Das and Alagirusamy [93]. These types of clothing comfort are psychological, thermo-physiological, sensorial/tactile aspects, and fitting comforts.

On the other hand, Li and Wong discussed comfort in several ways. As they explained, comfort relates to the subjective perception of various sensations and involves many aspects of human senses such as, visual, thermal, pain and touch [94]. Aspects of clothing comfort can be also classified into two broad categories namely, sensorial and non-sensorial comfort [95]. According to Slater, there are three types of comfort, namely, physiological, psychological and physical comforts [77]. Physiological comfort relates to the function of the human body, whereas physical comfort deals with the effect of environmental factors on the body. These comfort aspects are discussed in the context of apparel fabric which is made from the blend of synthetic and natural fibers. In the spite of this, the synthetic fiber negatively affects the wearer’s thermo-physiological comfort. In this regard, sensorial/tactile and thermo-physiological aspects of comfort are described in the subsequent sections.

### 2.2.2 Sensorial/tactile comfort

Sensorial/tactile comfort of clothing is the outcome of mechanical contact between the wearing garment and human skin along with the coordination of sensory systems at specified environmental conditions [93, 96, 97]. Sensorial aspect of comfort involves various sensations that affect the comfort of the wearer adversely when the dressing garment touches the skin fully or partly [98, 99]. Similarly, Li also stated that sensory comfort is as clothing interacts with the human body dynamically and continuously, it is directly in touch with the skin at the time of wear, which stimulates thermal, mechanical and visual sensations [100]. Typically, skin receptors, thermoreceptors, nociceptors and mechanoreceptors are stimulated by skin-clothing interaction during activity. Particularly, mechanical sensory perceptions of clothing mainly include roughness and scratchiness; dynamics of wear sensation; prickle, itch and rashes; and touch and pressure sensations. These sensory perceptions correlate mostly with the pain receptors which are found in the skin and associate largely to the surface and mechanical properties of the fabrics such as, cross section of protrude fiber ends, areal density, thickness and the smoothness of its surface [89, 93].

By touching the cloth, consumers can sense tactile sensations such as prickliness, roughness, stickiness, smoothness, stiffness, softness and scratchiness, whereas thermal sensations include coolness, breathability, warmth, chilliness and hotness [101]. Another perception is a moisture sensation and it consists of clamminess, wetness, stickiness, dampness, clingy and non-absorbent among others. And also, acoustic (sound and hearing) and odor characteristics as well as aesthetic perceptions are other sensorial properties [102]. On the other hand, looseness, heaviness, snugness, lightweight, stiffness and softness are included in pressure sensations which are related to body fit. Mechanical characteristics, fabric bulk and overall fit of the clothing garment to the body would have contributed to the pressure comfort. It indicates that the fabric properties can determine the sensorial comfort of clothing.

Sensorial comfort of clothing depends on the overall characteristics of fibers, yarns, fabrics and finishing processes as well as the skin of the wearer body's. Fiber characteristics deal with fineness, friction property, cross-sectional shape, moisture properties, resilience, length, compressibility and molecular orientation, while yarn type incorporates continuous filament, staple fiber, count, textured, twist and so on. And, fabric construction encompasses also woven, non-woven, knitted, thickness, weight, roughness and surface properties of fabrics. Furthermore, the finishing process includes the chemical and mechanical treatments of textile materials for apparel application [97, 103]. Some of the following sensorial comfort properties have been given more attention in this study.

#### i. Fabric hand Property

Fabric hand is a major aspect of sensorial comfort and it is used to assess the sensational property of fabric in the value chains of textiles, from producers to merchandisers, proposed for apparel use [102]. Fabric hand is mainly related to the low stress mechanical and surface properties of fabrics. It includes tensile, bending, shearing, compression and thickness properties. These properties provide mechanical stimuli to the body which is perceived by skin receptors [104]. Particularly, tensile performance of a fabric is a result of its structural and geometrical configuration. Mechanics of fibers are also linearly proportional with its length and inversely proportional to its diameter. Thus, fabric involved microfibers simply deform under pressure, moderately mimicking the body action. In order to dissipate high amounts of heat from the body, protrude and finer fibers with higher contact areas can produce cool sensations [76, 105]. Hang and drape of garment depends on the bending response of fabrics. Bending is affected by yarn bending behavior, construction of fabric and finishing treatment. Bending characteristics of textile can be expressed by using bending rigidities and bending hysteresis. In this regard, the viscoelastic properties of fibers and/or fabrics would be significant in the case of curvature bending. Typical finishing with enzyme treatments incorporates fibrillation and disintegration, then the fabric having a cooling nature often attended by a sensation of low bending rigidity and smoothness [103].

The shearing property of a fabric depends on its deformation, mobility of threads at intersection points of woven, yarn diameter and surface properties. Low shear rigidity would be the better fabric handle [106]. In this respect, loosely woven fabric gives simple shear deformation due to minimum yarn-to-yarn friction compared to closely woven fabrics. In knitted fabrics, the more stitch density develops the more stiffness [107]. Drape is another attribute of clothing and it depends on shearing and bending rigidities as well as mass of fabric. Appreciable draping of garment provides a comfortable perception, both psychologically and physically [97, 108, 109].

As far as handling is concerned, thickness and compression of fabrics are significant aspects. In this regard, the perception of softness is directly associated with the compression property of fabric. A perception of warm-cool sensation is subjected to the description of fabric hand [101]. On the other hand, smoothness increases the surface area of contact, whereas roughness decreases the effective area of skin-fabric contacts and thus, facilitates heat flow, which offers a cool hand. This significantly affects the comfort perception of warm-cool feeling in the underwear garments [110, 111].

#### ii. Electro-physical property

Although textile materials are protecting the body from harmful conditions, unfavorable effects are still existing due to the continuous contact of human skin with them. These effects comprise sticking of fabric to the skin, cleaning problems, dirt attraction and enhance the tendency of fiber ends to roll-up on their surface. This is the result of static electricity and directly affects the comfort of clothing fabrics. Static electricity in a charged fabric leads to clinging the wearing garment to the body or other materials, and forming unpleasant sensation to the wearer [112, 113].

Human body is an electrical conductor, but it can accumulate static electric charges when isolated from the ground. As it is shown in Figure 2.6, human dry skin is charged positively when rubbing with PET fabric. Since PET is highly sensitive to static electricity. it is also charged negatively. Static electricity is locally accumulation of either negative or positive charges and it is caused by the transfer of electrons from one object to another during pressurizing or rubbing contact of textile materials against other surfaces followed by separations. In addition to the amount of electrical charge, there are other two measurable parameters i.e. plus or minus signs and rate of decay or how fast it is dissipated.

The development of electrical charge depends on the fiber or fabric surface area of contact, the position of electrostatic series and electrical conductivity when textiles come into contact with or rub against surfaces such as human skin, other textiles, leather, metals and other materials. Additionally, relative humidity and temperature of the surrounding environment determine the generation and dissipation of static electric charges [114, 115].

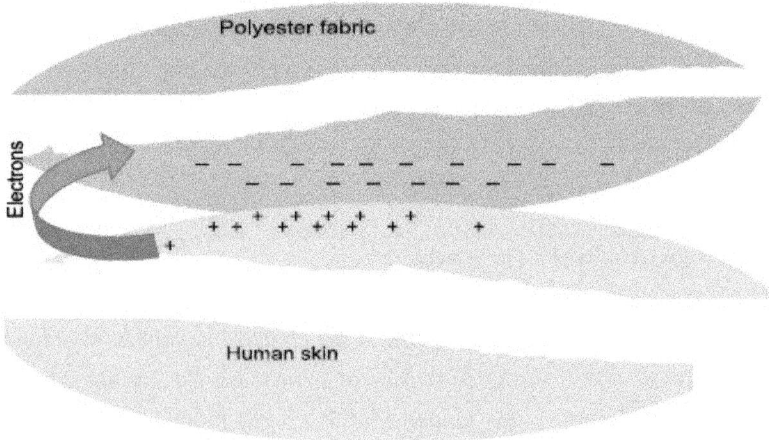


**Figure 2.6.** Static electricity during rubbing of human skin with PET fabric.

In general, most textile materials are electrical insulators and can build up charges readily during rubbing. In the dry/summer season, there is sparking when the clothes wear off at night. Various aspects of static-related health problems occur, under certain conditions, when the spark associating the wear off charged cloth can bring to development of dermatitis. Particularly, if a painful spark is frequently felt in the same skin area during clothing changes, a skin lesion may gradually develop in that area. These problems of clothing, in addition to sensorial comfort, also affect the psychological and physiological comfort of the wearer [116, 117]. Therefore, the effect of static electricity on sensorial comfort can be considered as one of the most important factors in clothing comfort.

### 2.2.3 Thermo-physiological comfort

Comfort is not only the function of psychological and sensorial aspects, but also must be understood in the perspective of thermo-physiological responses of the human body [90]. As Malik defined that the thermo-physiological comfort is the achievement of a comfortable wetness and thermal states via the passage of moisture and heat through a clothing fabric [118]. Kothari also stated that the thermo-physiological process involves the thermal comfort of the body, as a result of thermoregulatory responses together with the interaction of clothing and atmospheric conditions [4, 94]. The dynamic interactions of the human body with the environmental factors may lead to discomfort when it is beyond the limits of optimum level. Hence, the atmospheric factors like humidity, air movement, air and radiant temperature combined with the metabolic process of the human body can determine human-thermal environments [119]. The body mechanisms, for instance, thermoregulation, pulmonary, cardiovascular, skeletomuscular, digestive and central nervous systems, affect physiological comfort [120].

Clothing systems have a great contribution to the achievement of thermal-equilibrium by moderating the exchange of heat and moisture between the human body and the environment. Thus, thermal balance basically depends on the properties of fabric, personal factors and environmental variables [105, 121–124]. In fact, temperature, humidity and air velocity are the major environmental factors, whereas level of activity, clothing insulation and metabolic reactions are included in the personal factors. In addition, the fabric attributes and the design of garment belongs to the clothing properties [97].

Neurophysiological processes and human skin play a vital role in thermo-physiological comfort. Neurophysiology is a combination of physiology, which is the study of the sum of the body's parts and how they interrelate, and neurology is the study of the human brain and its functions along with the peripheral nervous system [119]. Neurophysiological processes are the neurophysiological mechanisms of the sensory reception systems of the body like skin, eyes, and other organs, when sensory signal impulses are stimulated from the interaction of the body with clothing and surrounding environments [94]. The interaction of human sensory organs with various stimuli can be presented sequentially as shown in Figure 2.7.

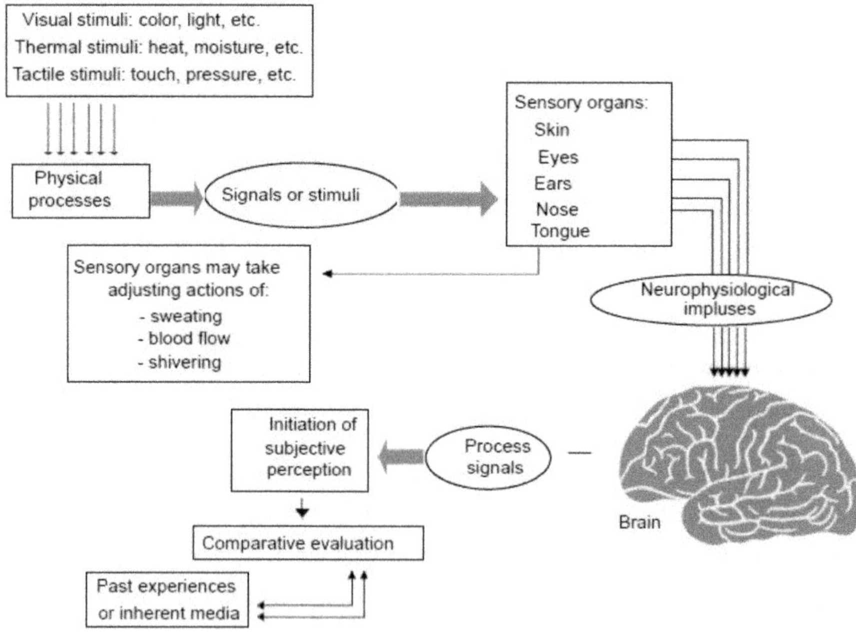


**Figure 2.7.** Simple illustration of human sensory mechanism, adapted from [120].

Additionally, the human brain can affect the physiological level of the body by changing the function of blood flow, sweating, shivering and others. Since the brain is the central component of the nervous system and contains neurons and supportive cells. Neuron cells also include dendrites, which receive information from adjacent cells and provide this to the cell body and axon, which assists the neuron to transmit this message. Therefore, it is necessary to understand the working mechanism of the brain and the sensory system to be capable of defining and understanding the neurophysiological comfort perception [126].

Human skin is the interface of a human body with its environment. It is made up of epidermis, dermis and hypodermis layers. The epidermis is the outer thinner layer and consisting of several layers of dead cells on top of a single living cell, whereas the dermis is the middle thicker layer and holding most of the nerve endings, receptor organs, sweat glands, fine muscle filaments and hair follicles. The receptor organs include thermoreceptors, Meissner's corpuscle, and nociceptors for the sensations of temperature, touch and pain, respectively. In addition, hypodermis is found under the dermis and contains layers of connective tissues and fat cells as well as Pacinian corpuscle for the sensation of pressure [94, 119, 126]. The three layers of skin and its sensory organs are depicted in Figure 2.8.

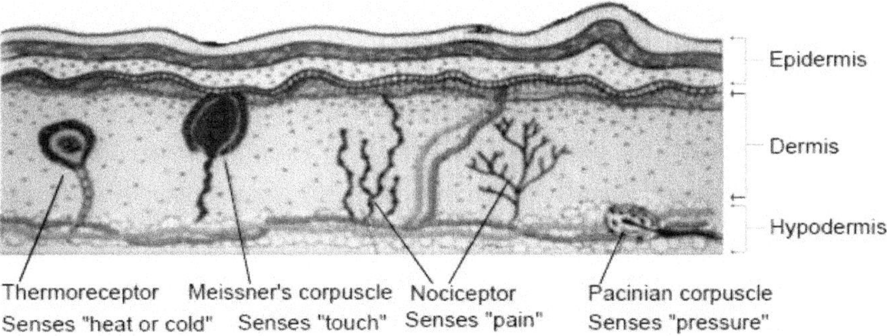


**Figure 2.8.** Skin structure and its sensory organs, adapted from [85].

The function of skin includes the regulation of temperature, protection, excretion, sensory perception and vitamin production. As pointed out earlier, sensation comprises the detection of various external stimuli such as mechanical contact with external objects, temperature changes through heat flow to or from the body surfaces, and traumatic and chemical induces. The receptor cells include photoreceptors (intensity and color of light), mechanoreceptors (sense of touch), thermoreceptors (heat and cold), nociceptors (sense of pain) and chemoreceptors (taste and smell) which react with the stimuli and the damaged tissues in order to take potential actions. Due to the stimuli, the receptor cells generate electrical activity and it is taken by means of unique nerve cells (conductors) to the brain. The signals send to the nerve cells (effectors) from the brain, which deliver the messages to the target groups like glands and muscles. These muscles and glands secrete hormones and substances in response to the signals. Finally, this phenomenon gives the actual feelings and sensations. In general, the skin is functioning as a dynamic system depending upon the requirement of the body [126, 127].

**Important parameters for thermo-physiological comfort**

**i. Thermal comfort**

In British standard–BS EN ISO 7730, thermal comfort is defined as a state of mind that articulates the satisfaction of a person with the thermal environment [128]. Although thermal comfort is used to express the response of the body, it is the satisfaction practiced by humans either consciously or unconsciously in a thermal state. A neutral or certain temperature thought to be suitable for one person may not be conducive to another person, similarly [129]. Since the metabolic rate, the body shape and size can influence thermal comfort. The percèption of warmness or coldness is caused mainly by the imbalance of the rate of heat dissipation from the skin surface and heat generation at the skin by metabolic processes.

The human body attempts to maintain a constant body's core temperature approximately at 37 °C. This core temperature slightly varies from person to person, but it should be maintained within a narrow limit. When the body becomes too cold, two processes are initiated by cold sensors. In the first process, vasoconstriction, reducing the blood flow near to the skin. Due to its fast cooling process, it may lead to limb injuries in outdoor conditions. However, the concentration of blood flows inside the body that provides heat to the necessary internal organs. The second process is shivering in order to increase internal heat generation by simulating muscles to be contracted. On the other hand, if the body also becomes too hot, the heat-sensors facilitate two mechanisms for body temperature regulation. In the first step, the blood vessels vasodilate, increasing the blood flow through the skin surface and consequently, initiates to start sweating. Sweating is an effective body cooling mechanism by releasing fluid (vapor and liquid) to the surrounding environment [130]. That means, the energy is taken from the skin to evaporate the sweat.

In general, the body temperature is regulated by the hypothalamus-sensors, which are cooling the body when the core temperature is greater than 37 °C, and skin sensors, which are warming the body when the skin temperature is below 33 – 34 °C. In order to maintain thermal comfort, two conditions must be fulfilled. The first one is that thermal neutrality by the actual combination of skin and body's core temperature, whereas the second condition is the body's energy balance, that means, the heat generated by metabolic should be equal to the heat lost from the body [4]. The body heat is transmitted in conduction, convection, radiation, evaporation and respiration. In the basic thermodynamic process, the heat exchange per unit body surface area (W/m$^2$) between the human and environment is expressed by the general heat balance equation:

$$M - W = C + C_k + C_{res} + R + E_{res} + E_{sk} \qquad (2.2)$$

Where M is metabolic rate (i.e. internal energy production); W is the external work; C is the heat loss by convection; $C_k$ is the heat loss by conduction; $C_{res}$ is the sensible heat loss due to respiration; R is the heat loss by radiation; $E_{res}$ is evaporative heat loss due to respiration; and $E_{sk}$ is the heat loss by evaporation from the skin, all are in W/m$^2$ unit. However, in most cases, the external work and loss of heat by conduction are neglected in the above equation. These are due to the small amount of the external work and the insignificant heat conduction through clothing from the body [93, 119, 131].

Heat can be transferred through clothing in the way of conduction, convection, radiation, and latent heat transfer by diffusion of water vapor and sweat (liquid) as shown in Figure 2.9. The first three heat transfer mechanisms are grouped as dry heat transfer and these are formed by the temperature difference between the skin surface and the environment [120]. In addition, the thermal insulation and evaporative resistance that determine the clothing effects on heat exchange by convection, radiation and evaporations. Thermal insulation is the resistance of clothing to transfer heat by convection and radiation, while evaporative resistance is the resistance of clothing to transfer heat by evaporation. Both of them consider the whole body surfaces, clothed and unclothed body parts, during heat exchange [131].

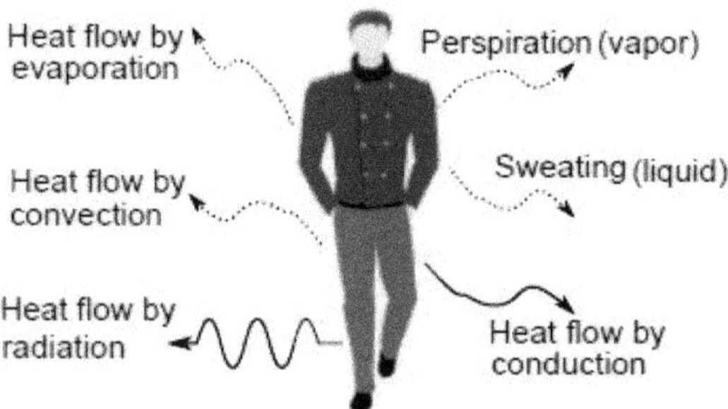


**Figure 2.9.** Heat transfer mechanisms through clothing, adapted from [9].

### ii. Moisture transmission

Transmission of moisture through clothes has a significant effect on the thermophysiological comfort of the body due to perspiration in the form of vapor and liquid. Under the normal atmospheric conditions and body activity, the insensible perspiration is continuously carried out from the skin layers in the form of water vapor. When the body perspiration remains in the vapor form, it is relatively comfortable. On the other hand, sensible perspiration occurs in the form of liquid sweat at higher physical activity and/or hot environmental conditions. In this situation, the skin is wetted and it feels the perception of discomfort. In addition, when the moisture transfer rate is reduced through the fabric, the skin temperature is increased and the resulting heat stress. In contrast, if the moisture is accumulated in the inner layer of the fabric system, it will reduce the thermal insulation of the clothing and result in undesirable body heat loss [93, 132, 133].

For comfort, not only in hot and cold weather but also at normal and high activity levels, the permeability of clothing fabric to water vapor should be optimized to allow the escape of water vapor which is continuously being released from the skin. Therefore, the arts of designing a clothing fabric for moisture transmission can be achieved by engineering the fabrics. Here, breathable coating is a typical example. It can be permeable to water vapor but inhibit the liquid water penetration and air permeability [4].

Absorption-desorption, diffusion and convection of water vapor together with wetting and wicking of liquid perspiration play an important role to maintain thermophysiological comfort. Liquid transmission through fabric is due to the fiber-water molecular attractions at the surface of the fibers by surface tension and effective capillary pore distributions [93, 118, 134]. The liquid transfers through a pore initially by wetting and subsequently with wicking. In the wetting process, the fiber-air interface is displaced with a fiber-liquid interface, whereas in the wicking process the liquid is forced by capillary pressure, which is produced when the liquid reaches the space between the fibers, and is pulled out along the capillary [135]. In the diffusion process, the vapor pressure gradient performs as the driving force to transfer moisture through the fabric and it depends on the air permeability of the fabric. Moreover, the sorption-desorption process uses to maintain a constant vapor concentration in the clothing fabrics by using the absorbing fabric as a source of moisture [136]. Forced convection also assists the moisture vapor transmission by flowing air over the moisture layer [137]. These all processes also depend upon the moisture content of the fabric, the type of fiber used, the perspiration rate and the environmental conditions like temperature, humidity and air velocity [93].

### iii. Air permeability

Air permeability is a measure of how well air is able to flow through a fabric. It can be measured on either dry or damp fabrics. A fabric which has good air permeability, however, does not necessarily have good moisture vapor permeability. Air permeability is likely to be lower in fabrics where the absorption of water leads to swelling of the fiber and the yarn.

### 2.2.4 Interaction of human-clothing-environment system

Clothing is an integral part of human life and it has a decorative, status, modesty and protective functions. However, the primary role of clothing is providing a layer of barriers to protect the body against drastic environments like radiation, electricity, chemical and other toxic substances. Especially environmental factors such as temperature, humidity and air movement can have a strong effect on the cooling and heating sensations of the exposed skin [94]. Without clothing, a person can live comfortably only in a narrow range of 26 to 30 °C [138], while with clothing, human beings can live and perform different activities in the wide range of -40 to 40 °C. Most of the time in daily life, more than 80 to 90% the human body is covered with clothing [139]. Therefore, understanding the interaction of the human-clothing-environment system is the most critical by considering it as an open system. The relationships among clothing, human body and environments are illustrated in Figure 2.10.

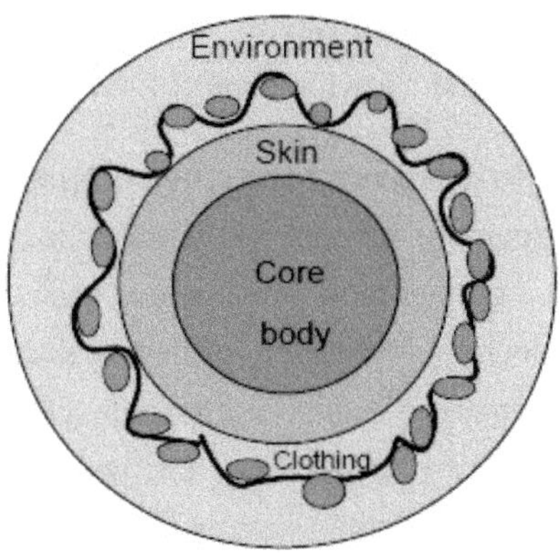


**Figure 2.10.** The human-clothing-environment system, adapted from [139, 140].

In the human-clothing-environment system, a clothing fabric provides a microclimate between the external environment and the naked body. It is characterized by thermal insulation, air permeability, thickness, weight, wind resistance and surface area. The body reacts to this microclimate by gaining and/or losing heat through conduction, convection, radiation and by evaporation of moisture in the form of perspiration and sweating. In the continuously heated body, a dynamic equilibrium is maintained when the body temperature is greater than skin temperature as well as the clothing surface temperature is less than the skin temperature and greater than the ambient temperature. Environment can be also characterized by relative humidity, temperature, air movement and radiant heat [97, 139, 141].

As it is depicted in Figure 2.10, the human body is denoted by two concentric circles, with the inner as the core body and the outer as the skin. The clothing forms another layer next to the skin, whereas the ambient air is found around the cloth. The layer of air also occurs between the skin and the clothing due to convection. The clothing system is considered as a single layer and that contains fibers, air and water vapor. The water vapor is diffused through the void space of fibers by the absorption-desorption process with the exchange of heat energy. The heat and moisture transfer processes are coupled by heat exchanges during phase change such as evaporation-condensation, absorption-desorption and freezing-melting. These processes determine the thermal function of clothing [139].

However, different climatic conditions need different types of clothing. In hot and dry climates, the temperature of atmospheric condition is in the range of 35-39 °C and sometimes it may be greater than 45 °C. In this case, the body would have a chance to receive heat energy in addition to the metabolic heat gain due to the negative temperature difference. Such a situation, clothing has to allow the transmission of vapor in order to decrease the heat strain by supporting the buffering effect [142]. Otherwise, the sweat would condense and accumulate on the skin and cause discomfort. Therefore, in hot and dry climatic conditions, hygroscopic fibers have good ventilation properties.

On the other hand, in hot and humid climates, atmospheric temperature and humidity are comparatively very high. In particular, the relative humidity is approximately 70-75% [143]. In this mild climate condition, transport of water vapor by convection is much faster than to transport it by diffusion in clothing due to the low vapor pressure gradient between the skin and atmosphere. In this condition, the fabric used next to the skin should immediately transfer the sweat to the outer layer of the fabric by wicking action. In addition, the fabric construction should allow the air circulation continuously over the skin in order to take away the moisture from the skin and keep it cool [144]. In the cold weather, a fabric should have high thermal insulation and effective sweat transfer mechanisms. The insulation property of clothing is the key factor to maintain the thermal balance of the body [131, 145].

### 2.2.5 Comfort properties of blend fabric

The characteristics of fabric determine the clothing comfort along with the wearer's attitude [90]. Clothing comfort is affected by the physical and chemical properties of fiber/yarn type and content. Indeed, a single type of fiber does not fulfill all the desired properties of textiles. To get the desirable properties in a fabric, blending natural and synthetic fibers is the most common technique. Some popular blend fabrics are P/C, wool/polyester, wool/acrylic, and polyester/cotton/viscose. Among these blends, the P/C blend fabric was used in this study. Because it is the most popular blend fabric and has been used for many applications [24], such as sportswear [25] and medical protective clothing after finishing treatment [26].

The proportion of cotton in the P/C blend fabric determines the comfortability of its blend fabric. Since, the cotton fiber provides wear comfort due to the presence of hydroxyl groups in its cellulose structure that imparts hydrophilicity to the fibers which leads to the higher moisture absorption and wettability [146]. Although the polyester fiber contributes high thermal stability, good strength, dimensional stability, good chemical resistance and acceptable easy-care properties to the blend fabric [147], the hydrophobic surface of polyester fiber affects the comfort properties of the blend fabric [148]. Particularly, the thermal comfort of the blend fabric depends on the abilities of heat and moisture vapor transmission as well as the absorption of the body sweats. In the case of the tropical climate and high level of physical activity, the sensible perspiration occurs in the form of liquid, namely sweat. Consequently, the skin is wet and feels discomfort [4, 149]. Similarly, when wearing a garment made from P/C blend fabric with a higher proportion of polyester, it tends to hold the perspiration in between the skin and the cloth due to the low water vapor absorbability of polyester compared to cotton [150].

As is already known, the hydrophobic nature of PET fiber has also a potential to develop electrical resistivity in the fabric, which is due to the tendency of synthetic fibers to static electrification. The build-up static charge, under low humidity conditions, does not dissipate for a long time because of the low conductivity of textile fibers. Consequently, accumulated static charges adversely affect the comfort properties of the garments when it rubs against the wearer's body, and at the same time, dust particles are more attracted towards the fabric surfaces [17]. Therefore, in order to modify the hydrophobic surface of PET fiber in the P/C blend fabric, finishing treatment is needed.

In this regard, several surface modification techniques have been applied on the fabric, aiming to the enhancement of its surface to be more hydrophilic. These methods include alkaline hydrolysis [147], aminolysis [151], applying natural biopolymer [152], gas treatment [153], silanization, UV treatment, flame treatment, and tethering molecules in bioconjugation [44]. Various P/C blend fabric surface modification techniques have been also used to develop better hydrophilic property such as applying microcrystalline cellulose particles as coating materials [150], enzymatic finishing with cutinase and lipase [154], corona discharge with chitosan [155], alkaline hydrolysis [147], and combining plasma treatment with oxidative in-situ chemical polymerization of pyrrole [156].

However, the surface modification through wet chemical processes is non-specific, changing the bulk property, consuming high amounts of water and energy, and releases hazardous effluent to the environment [157]. In addition, flame and UV treatment affects the optical properties of the polymer [44]. To resolve these problems, cold plasma surface modification is an ideal option as it is explained in detail previously in section 2.1.

Recently, a few studies have been performed by means of plasma surface modification to improve the comfort properties of certain textile fibers/fabrics. Particularly, some of the comfort properties of polyamide [81], wool [158], polyester [88, 159], bamboo viscose [84] and cotton [86, 87, 160] fibers have been enhanced after plasma treatments. However, not only a few studies have been done to this topic but also the thermal and tactile/sensorial comfort properties of P/C blend fabric have not been investigated yet via plasma surface modification.

In this study, the effect of low pressure plasma treatment on thermal and tactile/sensorial comfort of P/C blend fabric, such as wickability, thermal resistance, air permeability, water vapor permeability, hand feel and electro physical properties were investigated. In addition, the surface morphology and surface chemistry of the blend fabric were studied by using the surface characterization techniques.

### 2.2.6 Assessment of comfort

Since Peirce's well-known study in 1930, a number of research have been conducted to investigate the clothing comfort-related aspects [161]. According to Kilinc and Elmogahzy, the assessment of comfort can be classified into three categories viz. objective analysis (quantitative measures of tactile and thermal parameters), subjective analysis (psychological evaluations by using surveys, ratings and scales) and correspondence analysis by combining these two categories [162]. An objective measurement is performed by using different instruments, whereas subjective measurement is conducted with the opinion of the person that is directly participating in the evaluation of clothing comfort. Indeed, the objective measurement can be done accurately and precisely compared to a subjective one [77]. Some comfort assessment methods and techniques are stated below.

**Wear trial techniques**

Humans frequently use their hands to evaluate the tactile comfort of textile products. However, much of the tactile sensations of clothing comes from the parts of the body in wear situations. A wear trial is carried out under normal situations in large organizations usually by considering their own employees as an end-user. It often compares a new product against the previous one which is already satisfactorily in service such as low-priced common clothing articles. On the other hand, wear trial techniques are not widely applied in industry due to a number constraint. Some of these are its difficulty to control and organize, expensiveness, not possible in extreme conditions and takes a very long time when compared to the laboratory test. In fact, wear trial technique is a subjective measurement [94, 110].

**Thermal manikin**

Thermal manikin is an anatomical model of the human body which is constructed from a thermally conductive carbon-epoxy composite with embedded heat resistance wire and sensor elements. It also contains an independently heating/sweating controller, data measurement and analyzing subsystems.

Nowadays, thermal/sweating manikin is mainly used to determine the heat transfer characteristics of clothing and assesses the impact of thermal environment such as heat, flame and shock on the human body. Due to the development of digital systems, it is widely used in the mass production of textiles and clothing research centers in order to analyze the thermal interface of the human body with its environment. Since the thermal manikin measures heat losses quickly, accurately and reliably according to its standard [94, 163].

**Liquid moisture transfer**

The absorption of sweat, in the clothing-human skin microclimate, by the wearing garment and its transmission through and across the clothing can assist the perception of clothing comfort. Particularly, the way of moisture absorption at the inner portion of the fabric, transmission between the two sides and evaporation of it at the external surface considerably affects the wearer's comfort sensation. Since the moisture is a better heat conductor compared to the air. Therefore, in order to understand the phenomenon of moisture transmission in the fabric using Moisture Management Tester (MMT) is necessary. The MMT is used to measure the liquid transfer and moisture transport properties of the fabric. It is purposely designed to evaluate the moisture-management properties of the fabric. During measurement, the value of electrical resistance is changed based on the components of water and the water content in the fabric [94, 164].

**Objective measurements of sensorial/tactile comfort**

Sensorial comfort is directly correlated to the interactions of skin with fabric and is evaluated by the mechanical, structural and surface properties of the fabrics [102]. The objective evaluation of tactile or sensorial comfort includes measurement systems, partial analyzes and neurophysiological tests. The most widely used measurement systems are Kawabata Evaluation Systems for Fabrics (KES-F), Fabric Assurance by Simple Testing (Siro FAST) and Fabric Automated Modular and Optimization Universal System (FAMOUS). These techniques can quantify the fabric handle properties and predict the performance of fabric by realizing a series of objective analysis. In the partial analysis, the measurement of tensile and shear rigidity, bending rigidity, compressibility, surface-friction and profile roughness are conducted by using independent instruments [165]. Fabric Smart Tactile Tester (FSTT) is also developed in order to measure, record and analyze the thermal and the mechanical properties of the fabric in a single step and under the same atmospheric conditions [94].

Recently, Textile Softness Analyzer (TSA) and Textile electrode Tera-Ohmmeter (TO-3) have been developed to evaluate the hand-feel and electro-physical properties of the fabric, respectively [23]. The TSA is able to measure the three basic hand feel (HF) parameters individually, which are also felt by the human hand. These are softness (micro-surface variations), roughness (macro-surface variations), and in-plane stiffness of the fabric. The measurement is a two-step process. The first step is sound analysis for softness and roughness, whereas the second step is a deformation measurement for stiffness. From these three parameters and the number of plies, thickness, and weight of the fabric, an HF value can be calculated using a mathematical algorithm. The TSA measurement is minimal user influence, time and cost-saving, repeatable and reproducible HF values across a variety of grades, applications, and markets compared to the hand panel method [166, 167]. Regarding electrical resistivity measurement, two concentric ring electrodes of spacing suitable to the fabric being measured. With this type of instrument, the resistance is measured in both fabric length and width directions simultaneously compared to the parallel plate electrodes [168]. The TSA and TO-3 methods were also used in this study.

**Skin model**

The skin model is an internationally standardized thermoregulatory model of the human skin. It is used to investigate the thermophysiological wear comfort of clothing. The measuring unit of the skin model is made of sintered stainless steel. The introduced water in the channel under the measuring unit can evaporate through the pores of the plate as like that of sweating from the skin. Hence, heat and moisture transport in the measuring unit is comparable to those in the human skin. Different wear situations can be simulated with this skin model [92, 126]. The

Permetest is the simpler version of the Hohenstein Skin Model, and this study also used it. The working principle of the Permetest is illustrated in chapter three.

# CHAPTER THREE

# MATERIALS AND METHODS

## 3.1 Materials

### 3.1.1 Sample materials

In this dissertation work, two kinds of plain woven with a composition of 65% PET and 35% cotton blend fabrics, ready to make shirt cloth/workwear, were used as the substrate material. The substrate material is lightweight (115 g/m$^2$) P/C blend fabric (Carrington textiles, UK) which has 0.22 mm thickness, 144 ends per inch (44 Ne) and 72 picks per inch (44 Ne). The crimp (%) of wrap and weft yarn is 3.7 and 8%, respectively. This lightweight fabric was used for all tests like the wickability test, water vapor permeability, water vapor resistance, air permeability, electrical resistivity, fabric hand and thermal comfort properties as well as for surface chemistry and surface morphology analysis, except tensile strength and elongation test, before and after plasma treatment. Unfortunately, the lightweight fabric was used up before doing tensile strength and elongation test, meanwhile, Germany was under lockdown due to COVID-19 pandemic. Instead, the medium weight (150 g/m$^2$) P/C blend fabric (Klopman International, Düsseldorf, Germany) was also employed to test the tensile strength and elongation of the fabric. The count of warp and weft yarns for the medium weight blend fabric was 26 Ne, whereas the yarn densities were 105 ends per inch and 51 picks per inch. The size of each fabric sample depends on the dimension of the measuring equipment. The fabric samples were conditioned according to ASTM D 1776 for 24 h prior to plasma treatment and other measurements.

### 3.1.2 Chemicals

In this study, monomer, inert and reactive gases were applied for plasma surface modification of the P/C blend fabric. The atmospheric air and demineralized water vapor also used as the plasma gases. The purity and the molecular mass of the non-polymerizable gases specifically, argon (Ar), oxygen ($O_2$) and nitrogen ($N_2$)

(Westfalen AG, Germany) and the monomer HMDSO (Sigma-Aldrich Chemie GmbH, Germany) are given in Table 3.1.

**Table 3.1:** Purity and molecular mass of process chemicals/gases.

| Chemical/gas | Purity [%] | M [g/mol] |
|---|---|---|
| $O_2$ | ≥ 99.999 | 31.999 |
| $N_2$ | ≥ 99.999 | 28.013 |
| Ar | ≥ 99.999 | 39.948 |
| HMDSO ($C_6H_{18}OSi_2$) | ≥ 98.500 | 162.378 |

HMDSO [$(CH_3)_3SiOSi(CH_3)_3$] is a frequently preferred precursor as it is sufficiently volatile at ambient conditions, relatively non-toxic and non-flammable, and easily available from commercial sources. Its structure is depicted in Figure 3.1.

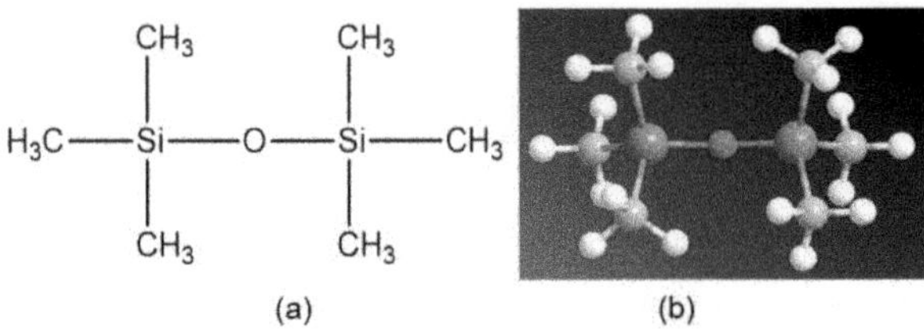


**Figure 3.1.** Chemical (a) and 3D (b) structures of HMDSO, adapted [169].

## 3.2 Plasma surface modification

### 3.2.1 Plasma reactor

The low-pressure plasma (LPP) reactor (Tetra 30 PC, Diener Electronic GmbH, Germany) set-up consists of a stainless steel vacuum chamber, a low-frequency (LF) generator, gas and monomer supply, and a pumping unit as schematically shown in Figure 3.2 [170]. A rectangular vacuum chamber (305 mm width, 300 mm height, and 370 mm depth) is double-walled to enable wall heating with a thermostat. This is important to prevent condensation of the precursor at the reactor wall. The volume of the vacuum chamber is approximately 34 L. It is enclosed in a hinged door that is made up of aluminum and glass pane. The surface modification of substrate is performed by a 40 kHz capacitively coupled LF-generator with maximum power of 1000 W. The LF-generator has automatic impedance adjustment since it is PC-interfaced. The LF powered floor electrode found in the chamber is made up of stainless steel and aluminum. The reactor wall acts as the grounded counter (secondary) electrode to the powered or primary electrode.

The movable bevelled floor tray with the dimension of 260 mm width and 335 mm depth is used for sample holders.

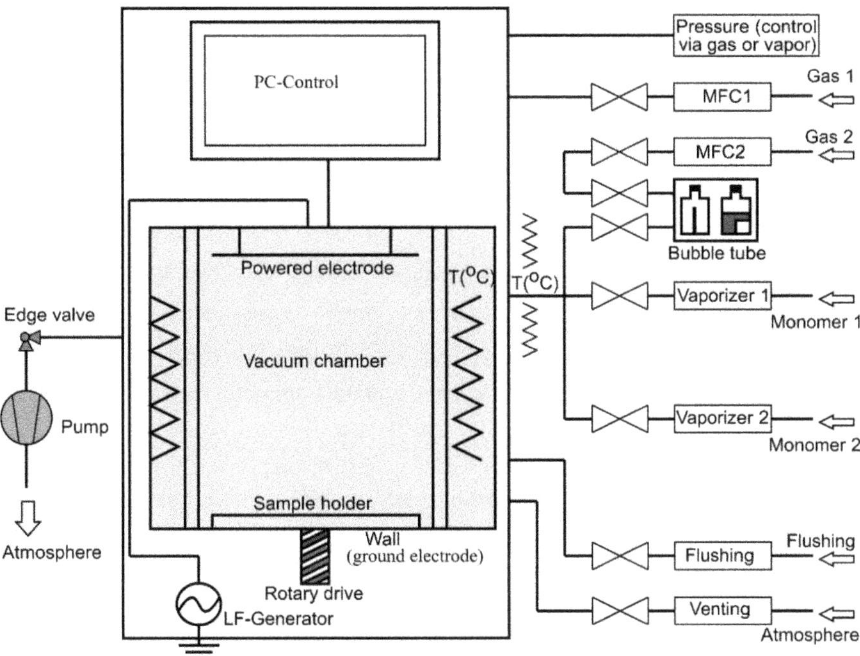


**Figure 3.2.** A schematic diagram of LPP reactor set-up, adapted from [170].

The non-polymerizable gases, such as oxygen, nitrogen and argon can be introduced into the chamber through two channels. The gas flow rate in gas 1 and 2 could be controlled by mass-flow-controller (MFC) 1 and 2, correspondingly, as shown in Figure 3.2. The MFC1 and MFC2 (MKS) can measure up to 500 and 1000 standard cubic centimeters per minute (sccm), respectively. The monomer 1 can be supplied to the chamber through vaporizer 1 in µL/min, whereas monomer 2 passes through vaporizer 2 with the help of cycle time and on time. The vaporizer is operated via pulse-width modulator. The measurement of pressure with the range of 0.01 to 10 mbar is controlled by a pirani sensor, but the pre-pressure at the glass bottle is approx. 0.5 to 1 bar. The dry vacuum pump (Leybold standard D16B and iQDP80 + QMB250, Edwards) with the suction power of 80 $m^3/h$ is a part of the reactor. All the processes including the pumping down, gas supply, plasma process, flushing, and venting are fully PC-controlled (window XP 7/8/10).

### 3.2.2 Plasma treatment

As it is previously illustrated in Figure 3.2, the LPP laboratory system has been used to treat the P/C blend fabric [170]. The samples were exposed to plasma modification by using different combinations of plasma parameters. In the case of oxygen plasma treatment, the plasma process was designed and analyzed by employing the Taguchi method to investigate the blend fabric's thermal comfort properties such as wickability, thermal resistance, air permeability, water vapor permeability, and water vapor resistance. In particular, this method was used to analyze the largest influential variables and optimal parameter levels of $O_2$–plasma treatment.

In this study, more attention was given to oxygen plasma. According to literature, $O_2$–plasma is most widely used to modify polymer surfaces [44]. It also has the potential to react with a wide range of polymers to form a variety of oxygen-containing functional groups, including C−O, C=O, O−C−O, and C−O−O [43]. Oxygen plasma provides activated oxygen in the form of ions (primarily $O_2^+$, $O^+$ and less amount of negative ions) and neutral atoms [171]. In $O_2$–plasma, the two kinds of processes i.e. etching of the polymer surface through the reactions of atomic oxygen with the surface carbon atoms, and the formation of oxygen-containing functional groups at the polymer surface through the reactions between the active species from the plasma zone and surface atoms, occur simultaneously. However, the parameters of a given experiment determine the balance of these two processes [43]. The experimental runs were set by using four factors and three levels, as presented in Table 3.2.

**Table 3.2:** The experimental runs of $O_2$ plasma with 4 factors and 3 levels.

| Runs | Flow rate (sccm*) | Pressure (mbar) | Power (W) | Time (min) |
|---|---|---|---|---|
| 1 | 40 | 0.2 | 600 | 5 |
| 2 | 40 | 0.4 | 700 | 10 |
| 3 | 40 | 0.5 | 800 | 20 |
| 4 | 50 | 0.2 | 700 | 20 |
| 5 | 50 | 0.4 | 800 | 5 |
| 6 | 50 | 0.5 | 600 | 10 |
| 7 | 60 | 0.2 | 800 | 10 |
| 8 | 60 | 0.4 | 600 | 20 |
| 9 | 60 | 0.5 | 700 | 5 |

* standard cubic centimeter per minute.

On the other hand, Air, Ar and $O_2$ were used as plasma gases, separately. The duration of the treatment time (5, 10 and 15 min) and operating power (200, 500 and 800 W) were varied in order to investigate the time and power effect of the plasma treatment on the electrical resistivity and hand-feel properties of the P/C blend fabric. The plasma gases were introduced into the vacuum chamber at a working pressure of 0.3 mbar and at a constant flow rate of 60 sccm, as well as at room temperature. At the end of each plasma process, air-flushing and venting were performed for 20 and 30 seconds, respectively [23].

Moreover, in order to evaluate the aging effect through wettability, the blend fabric samples were treated by using argon, air and oxygen plasmas, separately, at the condition of 500 W, 0.3 mbar, 10 min and 60 sccm. All plasma parameters, used in this study, were selected based on the recommendation of the plasma machine manufacturer to various types of materials, particularly for the Tetra 30 PC low-pressure plasma reactor [170]. Plasma treated and untreated samples were stored under standard condition for 1, 7 and 14 day/s. The plasma polymerization/coating was also performed by applying $O_2$/HMDSO, $N_2$/HMDSO, $H_2O$ /HMDSO and HMDSO-plasma in order to get durable surface treatment. In the plasma coating processes, firstly, the surface was cleaned and activated for 5 min by applying oxygen (50 sccm), nitrogen (50 sccm) and water vapor (liquid flow of 300 μL/min) plasmas, accordingly. Then, the plasma coatings were done at the condition of 700 W, 10 min and 0.3 mbar for $O_2$/HMDSO, $N_2$/HMDSO and $H_2O$ /HMDSO plasmas, whereas HMDSO plasma treatment was taken 15 min without surface cleaning and activation. The volumetric flow rate of HMDSO was 0.009 and 0.006 sccm for 10 and 15 min plasma coating treatment, respectively. A volumetric flow rate at standard conditions can translate into a specific mass flow rate by using equations of ideal gas law (3.1) and gas density (3.2) [172].

$$\mathrm{PV} = \mathrm{nRT} \qquad (3.1) \quad \text{and}$$

$$\rho = \frac{\mathrm{m}}{\mathrm{V}} \qquad (3.2)$$

Mass flow is equal to density times volumetric flow rate, as shown in equation (3.3).

$$\dot{m} = \rho . \dot{\mathrm{V}} \qquad (3.3)$$

Where,

P = Pressure [hPa][atm]

V = Volume [$cm^3$]

n = Number of molecules of gas [mole]

R = Universal gas constant [($cm^3$•atm)/(mole•K)]

T = Absolute temperature [K]

ρ = Gas density [g/$cm^3$]

m = Mass [g]

$\dot{m}$ = Mass flow [g/min]

$\dot{V}$ = Volumetric flow [$cm^3$/min]

$\dot{V}_S$ = Volumetric flow at standard conditions [$cm^3$/min].

In pulsed plasma systems, duty cycle (δ) is defined (Equation (3.4)) by the relative duration of the pulse on-time divided by the total time. Here, the pulse of vaporizer during plasma polymerization used the cycle time of 10 seconds (total time) and the due time of 25% ($T_{on}$). The 75% corresponding to afterglow duration ($T_{off}$) [169]. All plasma processes were carried out under the base pressure of 0.005 mbar and the pump pressure of 0.2 mbar at room temperature. In addition, all plasma treatments were fully automated.

$$\delta = \frac{T_{on}}{T_{off} + T_{on}} \qquad (3.4)$$

## 3.3 Comfort assessment methods

### 3.3.1 Permetest (Skin model)

The Permetest Skin Model (Sensora, Liberec, Czech Republic) was used to determine the relative water vapor permeability $P_{wv}$ (%), water vapor resistance $R_{et}$ ($m^2Pa/W$) and thermal resistance $R_t$ ($m^2K/W$) of the blend fabric samples. The measurement of the device was very similar to the ISO standard 11092 [173]. This skin simulator consists of the measuring head, semi-permeable layer, heating elements, fan, temperature and relative humidity sensors. The working principles of the Permetest is given diagrammatically in Figure 3.3.

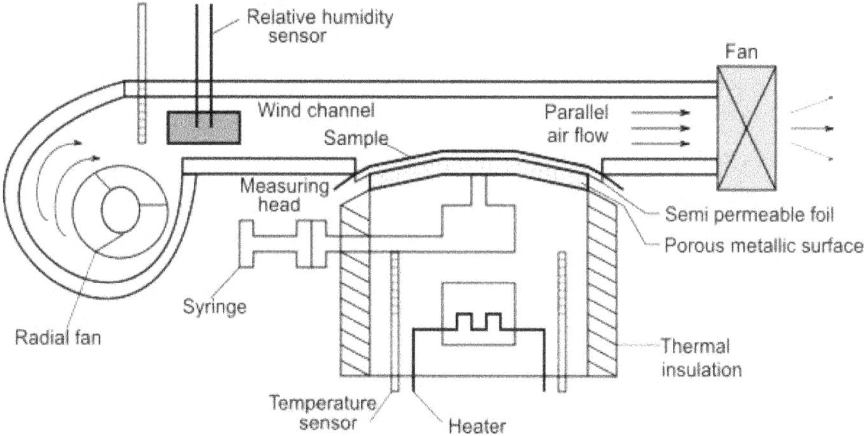


**Figure 3.3.** The working principle of Permetest, adapted from [120].

During the measurement of water vapor permeability and resistance, the measuring head was first masked with semi-permeable foil to retain the measured garment to be dry. Then, 10 mL of distilled water was poured through a special orifice under the measuring head. A little curved metallic porous surface along with a semi-permeable foil layer was moistened and exposed in a wind channel to a parallel air flow with the velocity of 2 m/s.

Special hydrophobic polypropylene reference fabric was used for calibration. The temperature of the measuring head was also kept at room temperature for isothermal working conditions. The Permetest was used to measure the amount of heat loss when evaporation of water escaped from the porous measuring head surface without ($u_o$) and with the fabric sample ($u_s$), respectively. Both values were used to calculate the mean value and standard deviations. The sizes of the samples were 11 x 11 $cm^2$, but the measured area was 8 cm in diameter. The mean of six readings has been taken for relative water vapor permeability and water vapor resistance tests. At the time of testing, the outer layer of sample was exposed to the air flow and the opposite side also overlapped to the porous humid surface, which was used to simulate the underwear garments filled with the liquid sweat.

The relative water vapor permeability Pwv (%) of the sample indicates the permeability of free measuring head surface and it can be calculated by using the following equation:

$$P_{wv}\,(\%) = 100\;\times\;u_s\;/\;u_o \qquad (3.5)$$

The water vapor resistance $R_{et}$ ($m^2Pa/W$) can also be expressed in the following equation:

$$R_{et} = (P_{wsat} - P_{wo})(1/u_s - 1\,/\,u_o) = C_r(100 - \varphi)\,(1/u_s - 1\,/\,u_o) \qquad (3.6)$$

Where $P_{wsat}$ and $P_{wo}$ are the values of water vapor saturation partial pressure in Pascal, valid for ambient temperature, and the actual water vapor partial pressure in the laboratory, respectively. In addition, $C_r$ is the value of the reference fabric for water vapor resistance (provided by the company as a standard) and φ is the relative humidity.

The thermal resistance $R_t$ ($mK.m^2/W$) of the dry blend fabric was measured by the Permetest. The whole procedure followed similar to that of water vapor resistance measurement, except the measuring head was kept dry. The temperature of the measuring head was kept 10 °C above the temperature of the air in the measuring channel. The thermal resistance $R_t$ has been worked out by the steady state electrical voltages with a fabric sample ($u_s$) and without a fabric sample ($u_o$) as well as by the sensitivity constant K (20.5 mK.m$^2$/W). It can be determined by using the following equation:

$$R_t = K(t_H - t_o)(1/u_s - 1 / u_o) \quad (3.7)$$

Where $t_H$ and $t_o$ are the temperature of measuring head and the temperature of the laboratory air, respectively.

### 3.3.2 Air permeability test

The air permeability of plasma treated and untreated blend fabric samples was tested on SDL ATLAS M021A air permeability tester at air pressure of 100 Pa and 20 cm$^2$ of testing area, in the standard conditions of 21 ±1 °C and 65 ±2% RH, according to EN ISO 9237. A specimen was clamped over the test head opening by pressing down the clamping arm which automatically started the vacuum pump. The pre-selected 100 Pa test pressure was maintained, and after a few seconds the air permeability of the test specimen was displayed in the pre-selected unit i.e. dm$^3$/s/m$^2$. An average of 10 readings was taken from each sample before and after oxygen plasma treatment.

### 3.3.3 Vertical wicking test

The effect of oxygen plasma treatment on the wickability of P/C blend fabric was evaluated by measuring the wicking height in the warp and weft directions according to DIN 53924 as depicted in Figure 3.4. The sample sizes of 250 x 30 mm$^2$ were cut along the warp and weft directions. A strip sample was vertically suspended on the clamp along with the plastic ruler and 15 mm of its lower end was immersed inside a reservoir of demineralized water with 1% C.I. Reactive Blue 220 (ALFA Industries, India) dye. The dye was used for tracking the movement of water against gravity on the strip. A spontaneous wicking occurs due to capillary force. An average of three readings of the wicking height was taken for each sample from the clamped ruler after 5 min wicking time.

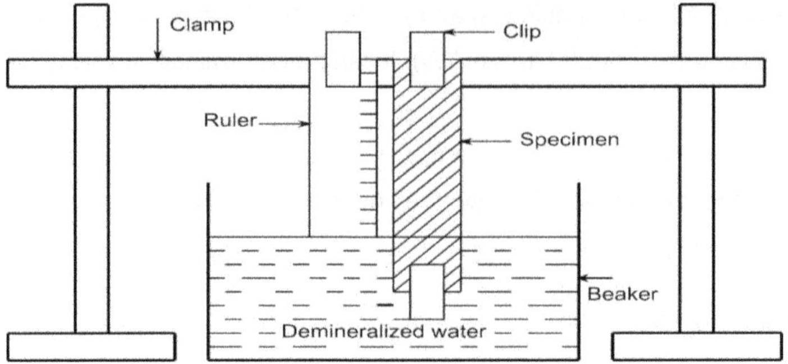


**Figure 3.4.** Vertical wicking apparatus.

### 3.3.4 Textile softness analyzer (TSA)

Softness, smoothness/roughness, stiffness and overall hand-feel (HF) value of the plasma treated and untreated P/C fabric samples were measured by TSA (Emtec Electronic GmbH, Germany) [174]. TSA was an objective measuring device for fabric hands and is capable of collecting all relevant parameters that had an influence on the haptic characteristics of P/C blend fabrics simultaneously. The haptic property was evaluated by measuring the sonic waves produced by rotating a set of blades on the clamped blend fabric surfaces, and at the same time, applying a constant 100 mN force on the surface. Thus, the vibration was captured by the microphone and recorded within a few seconds. Then, the blade applied 600 mN force, removed it, and again applied a 600 mN force using a constant loading rate. After this process, all variables and their values were accessible by Software "EMS Emtec Measurement system", which was interfaced with TSA.

In the resulting sonic spectrum, the fabric smoothness value (TS750) was given at 750 Hz signal peak by measuring the fabric vibration, as seen in Figure 3.5a, while the softness value (TS7) was obtained at 6500 Hz through the vibration of the rotating blade itself, as depicted in Figure 3.5b. The amplitude of acoustic signal peaks was measured in the dB unit. In addition, the stiffness value (D) was determined by measuring the sample deformation after applying a 600 mN force on it, as shown in Figure 3.5c. A combination of these parameters, with the blend fabric weight, thickness and a number of plies, were used to calculate the overall

HF value (HF = f (TS7, TS750, D, weight, thickness, number of plies)). In this experiment, only the bottom (back) side, i.e. the fabric side next to the skin, of the fabric sample was investigated and the average of the three specimens was taken from one set of tests and reported.

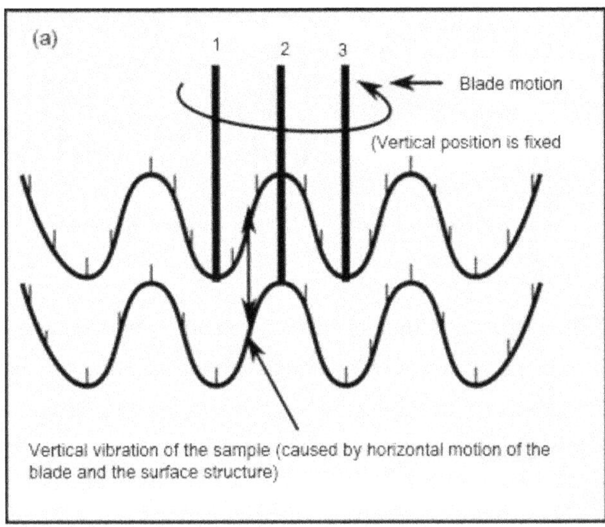


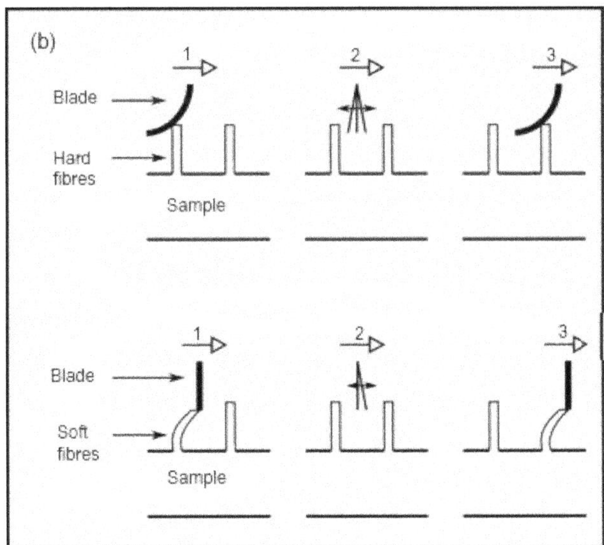

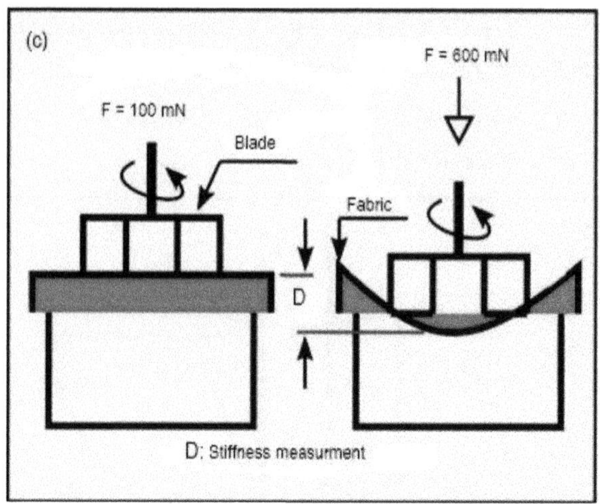


**Figure 3.5.** Technical principles of TSA by sound analysis and deformation measurement: (a) measuring of smoothness; (b) measuring of softness, and (c) measuring of stiffness ("Reproduced/adapted with the permission from [emtec Electronic GmbH, Leipzing, Germany], Alexander Grüner, Emtec TSA – Textile Softness Analyzer: A new and objective way to measure smoothness, softness and stiffness of textiles, emtec Electronic GmbH, 2018".) [174].

### 3.3.5 Tera-Ohmmeter (TO-3)

The Tera-Ohmmeter with a guard ring Textile Electrode (TO-3 TE 50, H. -P. Fischer Elektronik GmbH and Co. Industry and Laboratory Technology KG, Germany) was used to measure the surface and volume resistivities of the plasma treated and untreated P/C blend fabrics according to AATCC TM76 test standard [168], as shown below in Figure 3.6. Due to the presence of a concentric ring electrode, the resistivity of specimens was measured in both the length and width directions, simultaneously. During the surface resistivity measurement, the insulating disk was placed between the ground base plate and the specimen.

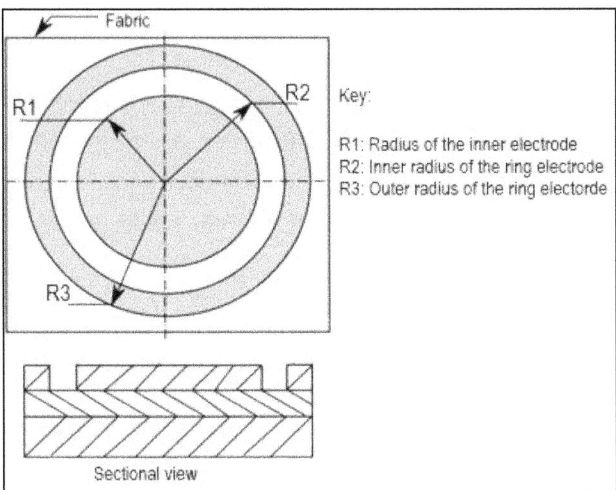


**Figure 3.6.** Surface and volume resistivities of ($\rho_s$, $\rho_v$) measurement configuration of concentric ring electrodes.

Finally, the ring electrode and inner electrode were positioned on the specimen, respectively, whereas in the volume resistivity measurement, the insulating disk was not required. Since the volume resistivity is the resistance to the leakage current through the whole body of non-conductive tested material under standard conditions. The resistivity was measured at 60 sec electrification time by applying 100 V. For each set of experiments, the mean of the three test specimens were taken [175].

## 3.4 TEGEWA drop test method and aging effect

TEGEWA drop test, similar to AATCC Test method 79, was used to determine the water absorbency of plasma treated and untreated P/C blend fabric. A size of 150 x 150 $mm^2$ samples and 2 g/L of patent blue V. dye solution were prepared. The sample was tensioned horizontally on the tensioning device without contact underneath. Then, the glass funnel was placed on the sample to hold the drop pipette at a distance of 40 mm. The amount of 0.05 mL ±10% solution was dropped from the pipette towards the sample once. The reading of sink-in time was started as soon as possible the drops landed on the sample. It was stopped when the bright and the shiny surface of the drop disappeared. The area of the color spot was measured with graph paper when the sample was completely dried. In addition, the aging effects of air, argon and oxygen plasma treated samples were investigated by conditioning them for 1, 7 and 14 days under standard conditions of 21 ±1 °C and 65 ±2% RH. The mean of the three readings was taken for each sample for analysis purposes.

## 3.5 Tensile strength testing

Fabric tensile tests were performed according to ISO 13934-1:2013 (Tensile properties of fabrics - Determination of maximum force and elongation at maximum force using the strip method). Tensile strength of the P/C blend fabrics were measured by Zwick Z2.5/TH1S machine (Zwick GmbH & Co. KG, Germany) under a standard condition of measurement. The test samples were prepared by initially cutting the blend fabrics in a size of 200 mm x 60 mm in weft and warp direction, and then yarns were removed from both sides to make a sample with 50 mm of final width. A pre-load of 2 N was applied to a test sample prior to testing. Grip to grip separation at the start position or the gauge length was 100 mm. A fabric specimen of 200 mm x 50 mm was extended at a constant rate of 100 mm/min until it ruptured. The average of five sample readings was recorded for maximum force, and elongation at maximum force using Test Xpert III software before and after plasma treatment.

## 3.6 Surface characterization techniques

### 3.6.1 Scanning Electron Microscopy

The scanning electron microscopy (SEM) with energy dispersive X-ray spectrometry (EDS) was used to investigate the surface morphology and surface chemistry of plasma treated and untreated P/C blend fabric. A SEM (Zeiss Sigma VP, Carl Zeiss Microscopy, Germany) allows to produce topography images of an object in the micro to nanometer range. When the non-conductive fabric samples were directly illuminated with an electron beam, the built-up negative charges on the sample surface affected the regular emission of secondary electrons. In order to prevent the charging effect, the surface of the samples was coated with gold powder by a sputter coater (Balzers UNION SD-300 Varian, OerlikonBalzers, Germany). A sputter coater applied 2-3 nm gold layer within 12 min in order to obtain a good conductive fiber surface prior to SEM imaging. The SEM images were recorded by means of a secondary electron and backscattered electron detector. The polyester and cotton fiber surfaces were scanned at working voltage of 5 kV and magnification of 10, 000X.

On the other hand, the EDS system (Oxford Instruments, Oxford, UK) associated with SEM was used to identify and quantify all elements in the P/C fabric samples except H, He, and Li. The X-ray detector is an OXFORD X-Max detector with 50 $mm^2$ detector size. The Gemini column is the area of the SEM where electrons were emitted, accelerated, bundled, focused, and deflected. The main functions of the Gemini® optics were the beam booster and an objective lens that consists of electrostatic and electromagnetic lenses together. The SEM was also equipped with a stub specimen holder, where the fabric samples were fixed by means of conductive stickers. The spot size of the primary electron beam determines the resolution of the instrument.

A focused electron beam rastered over the specimen under high vacuum and the emitted secondary electrons were collected by detectors. The secondary electrons were detected by an in-lens. The X-rays, secondary and backscattered electron signals provided the information about surface morphology and chemical composition of sample fabric. Consequently, the detected signal along with the beam position gave the data to develop a surface image. The operation of the SEM, EDX, optimization of the images, and handling of accessories controlled by SmartSEM® software.

In a variable pressure mode, images used an electron acceleration voltage of 20 kV and a 30 μm aperture size. The analyzed volume also corresponds to approximately 1 to 2 $\mu m^3$. However, the exact interaction of the electron beam in the material was not determined by the volume where X-rays were excited and information about the material was gathered depending on several material parameters like density, chemical composition and so on. Also the orientation of the sample to the detector can influence the count rate. As a result, the exact determination of the analyzed volume was quite difficult.

### 3.6.2 Fourier transform infrared spectroscopy

Fourier transform infrared spectroscopy (FTIR) is a common technique to estimate the functional groups of the states of matter. This technique is performed based on the interaction of infrared (IR) light with chemical bonds of a molecule. The incident light energy is absorbed and its intensity level is changed due to the excitation of vibrational modes of molecular bonds. Consequently, these vibrational modes produce specific frequencies for each chemical bond and functional groups. It can be used to characterize a substrate which may indicate the addition or omission of some functional groups due to plasma treatment.

In this work, the chemical composition of the P/C blend fabric samples were characterized by attenuated total reflection Fourier transform infrared spectroscopy (ATR-FTIR). ATR-FTIR analysis was carried out with Thermo Scientific Nicolet™ iS™ 10 FTIR spectrometer (Thermo Fisher Scientific Inc., Madison, WI, USA) powered by the OMNIC™ spectra™ Software. The FTIR spectrometer consists of an IR light source, interferometer, sample compartment, beam splitter and detector. Attenuated total reflectance (ATR), along with transmission, is one the most known sampling methods in FTIR spectroscopy. In ATR-FTIR spectroscopy, the ATR crystal is in contact with the sample. The incident IR light passes through the crystal and interacts with the sample surface coinciding with the ATR crystal, as shown in Figure 3.7a.

ATR is based on the total internal reflection when IR light interacts with a sample only at a point where IR light is reflected. On the other hand, transmission is based on IR light passing through the sample. The incident IR light from the spectrometer is directed onto the crystal at an angle (θ) so that the infrared light undergoes internal reflection. At each point of internal reflection an evanescent wave is formed when the light is absorbed by a sample. The evanescent wave of the infrared light has a limited penetration depth ($d_p$), as depicted in Figure 3.7b.

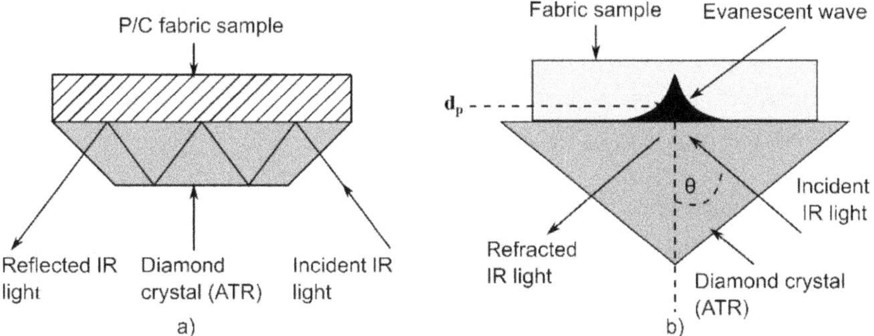


**Figure 3.7.** Working principle of ATR-FTIR at five-bounce (a) and single-bounce (b), adapted from [176].

In this study, a deuterated triglycine sulfate (DTGS) detector, helium–neon laser or He-Ne laser, KBr beam splitter with standard mid-infrared (IR), and a diamond crystal was used. The spectra were collected using 16 scans for each sample with a 4 $cm^{-1}$ spectral resolution between 650 and 4000 $cm^{-1}$. The spectra were recorded as transmittance ($T_{IR}$), defined as the intensity ratio of the transmitted to the incident light, and converted to absorbance ($A_{IR}$) according to the following equation:

$$A_{IR} = -log_{10}T_{IR} \quad (3.8)$$

## 3.7 Data analysis

Experimental runs and optimization of oxygen plasma processes, for thermal comfort properties, were designed and analyzed by Taguchi method (Minitab software, version 18). The Taguchi method is a special form of a statistical technique and mainly used by engineers and quality assurance professionals for process optimization and solving production problems [177]. In this design, variables are sorted according to their influence on the response variable. The quantitative relationship between the variables and responses were analyzed based on the signal-to-noise (S/N) ratio approach. A S/N ratio is a measure of robustness, which can be used to identify the control factor settings that minimize the effect of noise on the response. Signal represents the mean values of designating desirable components, while noise symbolizes the introducing undesirable components. For static designs, Minitab provides three alternatives to make analysis: nominal-is- best, smaller-is-better, and larger-is-better [177]. In all cases, the S/N ratio is to be maximized.

In this dissertation work, smaller-is-better and larger-is-better were used to analyze the responses. The S/N ratio of the smaller-is-better and larger-is-better were calculated for each factor level combination using equations (3.9) and (3.10), respectively.

$$S/N = -10log\left[\frac{(\sum_{i=1}^{n}(y_i)^2)}{n}\right] \qquad (3.9)\ and$$

$$S/N = -10log\left[\frac{(\sum_{i=1}^{n} 1/(y_i)^2)}{n}\right] \qquad (3.10)$$

Where y is the responses for the given factor level combination and n is number of responses in the factor level combination. For other statistical analysis, Origin software was used.

# CHAPTER FOUR

# RESULTS AND DISCUSSION

## 4.1 Influence of oxygen plasma on thermal comfort properties of P/C blend fabric

### 4.1.1 Water-vapor permeability and resistance of the fabric

After oxygen plasma treatment, the Permetest (Skin Model) was used to determine the relative water vapor permeability Pwv (%), water vapor resistance Ret ($m^2Pa/W$) and thermal resistance Rt ($m^2K/W$) of the blend fabrics. During the measurement of water vapor permeability and resistance, the measuring head was first masked with semi-permeable foil to retain the measured garment to be dry. Then, 10 ml of distilled water was inserted through a special orifice under the measuring head. A metallic porous surface along with a semi-permeable foil layer was moistened and exposed in a wind channel to a parallel air flow with the velocity of 2 m/s. The Permetest measured the amount of heat loss when evaporation of water escaped from the porous measuring head surface without and with the fabric sample, respectively. At the time of testing, the outer layer of sample was exposed to the air flow and the opposite side also overlapped to the porous humid surface, which was used to simulate the underwear garments filled with the liquid sweat. The relative water vapor permeability Pwv (%) of the sample indicates the permeability of free measuring head surface.

The bar graphs of water vapor permeability and water vapor resistance from oxygen plasma treated and untreated P/C blend fabrics are shown in Figure 4.1. The relative water vapor permeability $P_{wv}$ (%) of the plasma treated fabric increased in the range of 71.8 to 74.8% compared to 71.4% of untreated fabric. At experimental run 6, the highest water vapor permeability is achieved as depicted in Figure 4.1. The level combinations of all factors for run 6 (50 sccm, 0.5 mbar, 600 W & 10 min) is similar to the optimum parameters of water vapor permeability (50 sccm, 0.5 mbar, 600 W & 10 min) except pressure as shown in Table 4.1. In addition, the level of influcial factor i.e. flow rate of oxygen is the same as the value of oxygen flow rate in run 6 (50 sccm). This improvement is most likely related to the formation of cracks and grooves over the surface of blend fabric due to the etching and cleaning effect of oxygen plasma treatment [88]. These developed cracks and grooves on the surface of fibers would increase the capillary pressure and assist water vapor permeability [84]. On the other hand, water vapor resistance of plasma treated fabric decreased in the range of 1.73 to 1.98 $m^2Pa/W$ compared to 2.02 $m^2Pa/W$ of untreated blend fabric due to the cleaning effect of oxygen plasma. It is the reciprocal of water vapor permeability. Water vapor transmission through clothing provides better comfort both indoors and outdoors. If the clothing layers are impermeable, moisture is trapped between the skin and clothing and heat accumulates on the body. Consequently, the wearer feels discomfort [4].

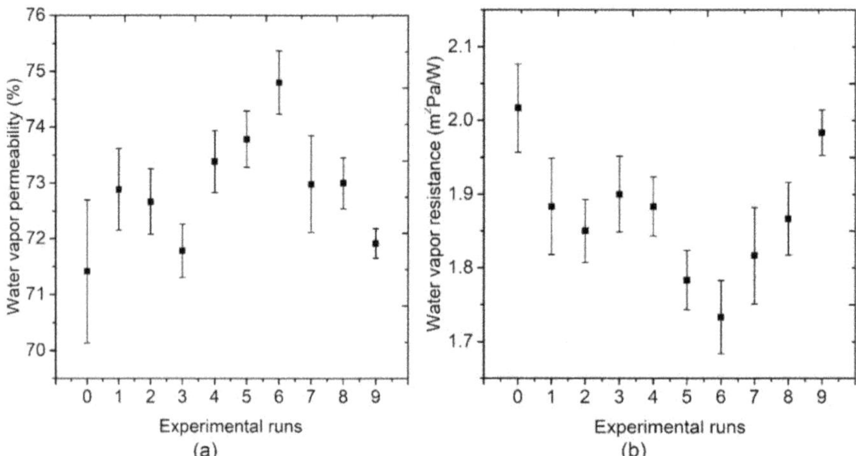


**Figure 4.1.** Oxygen plasma on water vapor permeability (a) and water vapor resistance (b) of the fabric.

In fact, surface topography and pore structure of the fiber surfaces have a significant influence on water vapor transmission of the fabric. As a result of this influence, plasma-treated P/C blend fabric had better comfort property by regulating water vapor that comes from the body than untreated fabric [86, 87]. Moreover, better comfort property could be achieved by increasing water vapor permeability and decreasing the water vapor resistance through oxygen plasma treatment for tropical climatic conditions. Otherwise, moisture is trapped between the skin and clothing which leads to the wet skin and discomfort [4].

Optimization of oxygen plasma process parameters for water vapor permeability and resistance of blend fabric was analyzed by Taguchi method. In this study, the plasma process primarily depends on the flow rate, discharge power, operating pressure and exposure time, since other parameters are constant. As could be seen from Table 4.1, the flow rate of oxygen brought the highest effect on water vapor permeability and water vapor resistance of plasma treated fabric. With respect to the response table of S/N ratio, the optimum parameter levels attained at 50 sccm, 0.4 mbar, 600 W and 10 min, by choosing "Larger is better", for water vapor permeability (S/N[a]), while for water vapor resistance (S/N[b]), it achieved that, by selecting "Smaller is better", at the levels of 60 sccm, 0.5 mbar, 800 W and 20 min. In warm conditions, releasing water vapor (body sweat) to the surrounding environment is necessary to maintain the body at a comfortable level. This is a reason for the selection of "Larger is better" and "Smaller is better" for water vapor permeability and water vapor resistance of plasma treated fabric, respectively.

**Table 4.1:** Response table for mean of S/N ratio to optimize water vapor permeability and (S/N[a]) and resistance (S/N[b]) of the fabric.

| Runs | Level | Flow rate (sccm) | Pressure (mbar) | Power (W) | Time (min) |
|---|---|---|---|---|---|
| **Average** | 1 | 37.20 | 37.27 | 37.33* | 37.25 |
| **S/N[a], dB** | 2 | 37.38* | 37.28* | 37.22 | 37.32* |
| | 3 | 37.22 | 37.24 | 37.24 | 37.23 |
| *Larger is* | **Delta** | 0.18 | 0.04 | 0.11 | 0.09 |
| *better* | **Rank** | 1 | 4 | 2 | 3 |
| **Average** | 1 | -5.490 | -5.416 | -5.252 | -5.504 |
| **S/N[b], dB** | 2 | -5.113 | -5.276 | -5.606* | -5.121 |
| | 3 | -5.534* | -5.446* | -5.280 | -5.512* |
| *Smaller is* | **Delta** | -0.421 | -0.171 | -0.353 | -0.391 |
| *better* | **Rank** | 1 | 4 | 3 | 2 |

*Optimum parameter level; [a]water-vapor permeability; [b]water-vapor resistance.

### 4.1.2 Thermal resistance of the fabric

Figure 4.2 shows the thermal resistance of plasma treated and untreated fabrics. According to the results, the plasma treated samples had lower insulation property compared to untreated samples. Particularly, the thermal resistance decreased at least by 20.16%. This means that oxygen plasma treatment reduced the thermal resistance of P/C blend fabric. Because plasma treated samples have more air permeability due to the formation of new pores by means of surface cleaning and etching effect [178]. Consequently, heat is escaped from the fabric using the movement of air and water vapor through convection mechanism. When the thermal resistance of the clothing material decreases in a hot climatic environment, the thermal energy is more easily dissipated and the body cools down. Therefore, this type of treatment can be an option to change the thermal resistance of produced fabric and/or garment. Thermal resistance of the blend fabric was governed by plasma conditions, types of fiber, and fabric construction [179, 180].

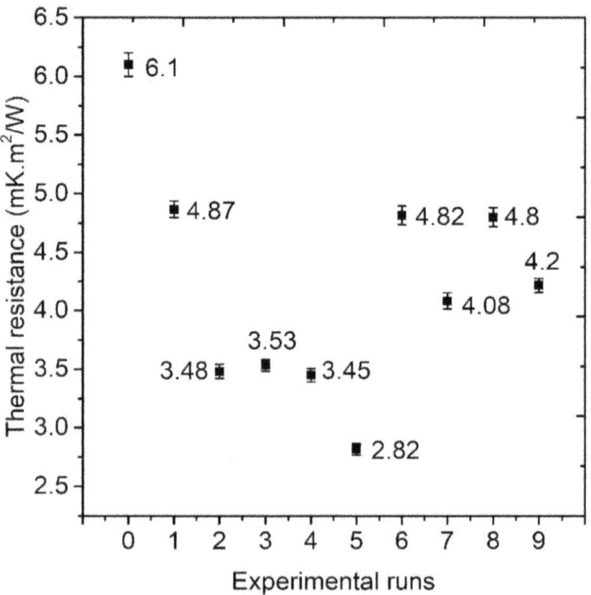


**Figure 4.2.** Effect of oxygen plasma on the thermal resistance of the fabric.

The optimum parameters and influential variables of blend fabric thermal resistance is indicated in Table 4.2. As per the delta value, discharge power on thermal resistance had the greatest effect in the oxygen plasma process. The optimum result of the thermal resistance (S/N) scored at 60 sccm, 0.5 mbar, 600 W and 10 min by selecting the "*Smaller is better*" during the analysis of the response with Taguchi method.

**Table 4.2:** Response table for mean of S/N ratio to optimize thermal resistance of the fabric.

| Runs | Level | Flow rate (sccm) | Pressure (mbar) | Power (W) | Time (min) |
|---|---|---|---|---|---|
| **Average S/N, dB** | 1 | -11.85 | -12.25 | -13.68* | -11.75 |
| | 2 | -11.14 | -11.16 | -11.37 | -12.24* |
| | 3 | -12.79* | -12.38* | -10.73 | -11.79 |
| *Smaller is better* | **Delta** | -1.65 | -1.22 | -2.95 | -0.49 |
| | **Rank** | 2 | 3 | 1 | 4 |

*Optimum parameter level

### 4.1.3 Air permeability of the fabric

The air permeability of plasma treated and untreated fabric is depicted in Figure 4.3. Regarding air permeability, two kinds of results were observed after plasma treatment. Specifically, the results revealed that the air permeability at experimental runs of 1 and 7 were lower than control samples, whereas others were higher than untreated samples.

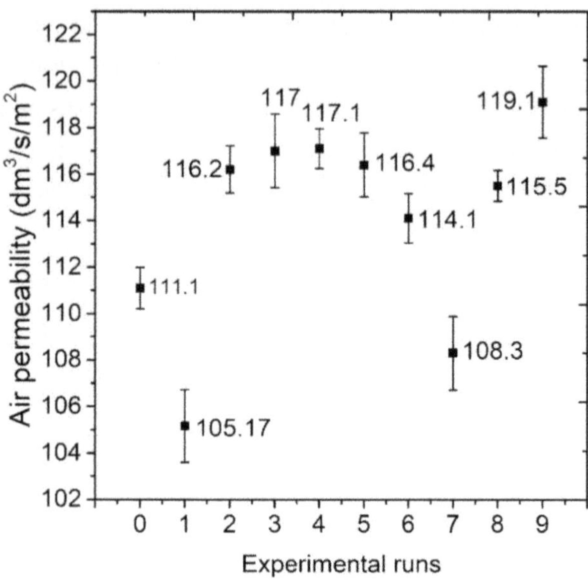


**Figure 4.3.** Influence of oxygen plasma on air permeability of the fabric.

Many reasons may be taken into consideration to explain it. However, the surface roughness of cotton fiber plays a great role to determine the air permeability. When the degree of roughness increases on cotton fiber, the air is trapped in and cannot escape easily from the fabric [84]. At a lower operating pressure (0.2 mbar) and shorter treatment time, the degree of roughness on cotton is higher than that of polyester. This may be the reason for experimental runs of 1 and 7. Because operating pressure had the greatest effect on the air permeability of the blend fabric according to Table 4.3. Surprisingly, previous research works have also shown different outcomes regarding air permeability of cotton fabric [181, 182].

In this study, the composition of P/C blend fabric is 65% polyester and 35% cotton. Hence, surface modification depends on the amorphicity and crystallinity of individual fiber. The formation of more voids and cracks on the cotton fiber surface probably would lead to the reduction of air permeability of the blend fabric like experimental runs of 1 and 7 [159]. However, the degree of roughness on the surface of polyester fiber is higher than the formation of voids. Probably, it facilitates the circulation of the air through the fabric rather than storing air in grooves like cotton [178]. Because plasma treated samples have more air permeability due to the formation of new pores by means of surface cleaning and etching effect. Therefore, the overall air permeability of the blend fabric mainly depends on the surface modification of polyester and cotton fibers.

Here, similar methods and factors were used like that of thermal resistance analysis. The optimum parameters and influential variables of blend fabric air permeability is indicated in Table 4.3. As per the delta values, operating pressure had the greatest influence on air permeability of the fabric in the oxygen plasma treatment process. The optimum outcomes of air permeability (S/N) found at 50 sccm, 0.5 mbar, 700 W and 20 min by choosing the "*Larger is better*" using Taguchi method.

**Table 4.3:** Response table for mean of S/N ratio to optimize air permeability of the fabric.

| Runs | Level | Flow rate (sccm) | Pressure (mbar) | Power (W) | Time (min) |
|---|---|---|---|---|---|
| **Average S/N, dB** | 1 | 41.02 | 40.82 | 40.93 | 41.08 |
| | 2 | 41.28* | 41.29 | 41.39* | 41.03 |
| | 3 | 41.14 | 41.33* | 41.11 | 41.32* |
| *Larger is better* | **Delta** | 0.26 | 0.51 | 0.45 | 0.29 |
| | **Rank** | 4 | 1 | 2 | 3 |

*Optimum parameter level

### 4.1.4 Wickability of the fabric

Wickability (vertical wicking) of oxygen plasma treated and untreated P/C blend fabrics are presented in Figure 4.4. By observing the bar graphs, it can be depicted that the wicking heights raised in the range of 57.3 to 64.3 mm as compared to 40 mm (0) in the warp (Figure 4.4(a)) and 52.3 to 60 mm as compared to 38 mm (0) in the weft (Figure 4.4(b)) directions after 5 min wicking time. That means, wickability of treated fabrics increased by the range of 43.25 to 60.75% and 37.63 to 57.89% compared to untreated fabric in the warp and weft directions, respectively. In plasma treated and untreated P/C blend fabrics, the wicking height in the warp became higher than that of weft directions due to the difference of yarn crimp percentage and twist. The crimp of weft yarn is higher than warp yarn. This is due to a higher warp yarn density (144 EPI) compared to weft yarn (72 PPI). The crimp (%) of warp and weft yarn is 3.7 and 8%, respectively. Therefore, the wicking process in the weft yarn needs more time than warp yarns. In addition, during weaving, the warp yarns are subjected to greater stress and usually have tighter twists than the weft yarns. Increasing yarn twist means increased the capillary spaces of the fabrics. This result depicts that crimp and twist can play a crucial role in wickability of the fabric [183–185].

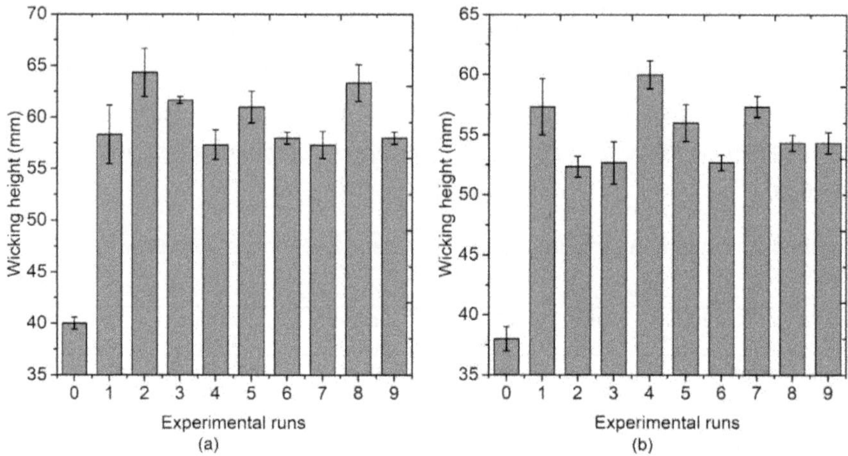


**Figure 4.4.** Effect of oxygen plasma on wickability of fabric in warp (a) and weft (b) directions.

Plasma etching altered the surface morphology of cotton and polyester fibers. After plasma bombardment, the hydrophobic layer of polyester fiber modified and also hydrophilicity improved due to introduction of polar groups. In addition, cracks, roughness and nano-voids correspondingly developed on the surface of both fibers. These kinds of surface morphology can increase the capillary pressure of the liquid. In other words, the movement of liquid through the fabric can be governed by capillary force [88]. Therefore, oxygen plasma treatment can improve the wickability of P/C blend fabric.

Optimization of the oxygen plasma process parameters for wickability of blend fabric was analyzed by Taguchi method. In this study, the plasma process primarily depends on flow rate, discharge power, operating pressure and exposure time. While, other parameters are kept constant. As can be shown from Table 4.4, the discharge pressure brought the largest influence on the wickability of the blend fabric both in warp and weft directions according to the delta values of S/N ratios, by selecting the "Larger is better", from Minitab 18 software®. The optimum parameter levels were found at 40 sccm, 0.4 mbar, 800 W and 20 min for the warp direction. Similarly, the optimum parameter levels occurred at 50 sccm, 0.2 mbar, 800 W and 5 min for the weft direction. As it was revealed in the optimum parameter levels for the weft direction, the wickability of the blend fabric is highly enhanced at low pressure due to a large number of molecular fragmentation in the gas phase. However, at a higher pressure, wickability is decreased remarkably due to the low activation rate, which leads to the production of a low amount of molecular fragmentation in the gas phase [186].

**Table 4.4:** Response table for mean of S/N ratio to optimize wickability of the fabric.

| Runs | Level | Flow rate (sccm) | Pressure (mbar) | Power (W) | Time (min) |
|---|---|---|---|---|---|
| **Average S/N[a], dB** | 1 | 35.72* | 35.16 | 35.51 | 35.41 |
| | 2 | 35.34 | 35.95* | 35.49 | 35.52 |
| | 3 | 35.48 | 35.43 | 35.53* | 35.61* |
| *Larger is better* | **Delta** | 0.38 | 0.79 | 0.05 | 0.20 |
| | **Rank** | 2 | 1 | 4 | 3 |
| **Average S/N[b], dB** | 1 | 34.63 | 35.25* | 34.75 | 34.88* |
| | 2 | 34.95* | 34.67 | 34.81 | 34.65 |
| | 3 | 34.81 | 34.47 | 34.84* | 34.86 |
| *Larger is better* | **Delta** | 0.31 | 0.78 | 0.09 | 0.23 |
| | **Rank** | 2 | 1 | 4 | 3 |

*Optimum level; [a]Wickability in warp direction; [b]Wickability in weft direction.

The results of One-way ANOVA for each thermal comfort property of oxygen plasma treated and untreated blend fabrics are summarized in Table 4.5. It clearly revealed that means of experimental runs for each thermal comfort property is significantly different ($p<0.05$) at 95% confidence level. Therefore, oxygen plasma treatment had a significant change on the comfort properties of P/C blend fabric.

**Table 4.5:** One-way ANOVA on thermal comfort properties of the P/C fabric.

| Source of variation | DF | Sum of Squares | Mean Square | F-Value | Prob>F |
|---|---|---|---|---|---|
| Between groups | 9 | 1247.87 | 138.65 | 19.26 | 5.44178E-8[a] |
| Within groups | 20 | 144 | 7.2 | | |
| Total | 29 | 1391.87 | | | |
| Between groups | 9 | 966.83 | 107.43 | 21.78 | 1.86507E-8[b] |
| Within groups | 20 | 98.67 | 4.93 | | |
| Total | 29 | 1065.5 | | | |
| Between groups | 9 | 54.57 | 6.06 | 2.17 | 0.04022[c] |
| Within groups | 50 | 139.69 | 2.79 | | |
| Total | 59 | 194.26 | | | |
| Between groups | 9 | 0.39 | 0.043 | 2.81 | 0.00947[d] |
| Within groups | 50 | 0.77 | 0.015 | | |
| Total | 59 | 1.16 | | | |
| Between groups | 9 | 49.44 | 5.49 | 190.30 | 0[e] |
| Within groups | 50 | 1.44 | 0.03 | | |
| Total | 59 | 50.88 | | | |
| Between groups | 9 | 1763.49 | 195.94 | 12.35 | 1.60083E-12[f] |
| Within groups | 90 | 1428.20 | 15.87 | | |
| Total | 99 | 3191.69 | | | |

[a]Wickability in warp direction; [b]Wickability in weft direction; [c]water vapor permeability; [d]water vapor resistance; [e]thermal resistance; [f]air permeability.

## 4.2 The effect of air, Ar and $O_2$ plasmas on sensorial comfort properties of P/C blend fabric

### 4.2.1 Fabric hand properties of plasma-treated fabric

Fabric hand properties, including stiffness, softness, smoothness and hand-feel (HF), are the fundamental comfort quality parameters of clothing fabric. However, different treatment processes sometimes affect these properties. Here, plasma treated P/C blend fabric hand properties were evaluated by TSA after being conditioned for 24h at 65 ± 2% RH and 21 ± 1 °C. During plasma treatment, an operating pressure of 0.3 mbar, gas flow rate of 60 sccm and at temperature of 23 ± 2 °C remained identical, while the discharge power and treatment time were varied from 200 to 800 W and 5 to 15 min, respectively. Here, only the effect of discharge power was discussed, due to the similar effect of power and treatment time on the fabric hand properties. However, when comparing the two parameters, the effect of discharge power on the fabric hand was slightly higher than the treatment time. The mean of three readings was taken on the back (bottom) side of the blend fabric for each datum in the figure. Because this side of the fabric was directed towards the wearer's body.

### Roughness

Roughness is correlated to fabric surface slipperiness, as a function of frictional property, whereas roughness is related to the surface texture of a fabric [187]. As TSA results depicted (Figure 4.5), the roughness value of all plasma treated samples were greater than untreated samples.

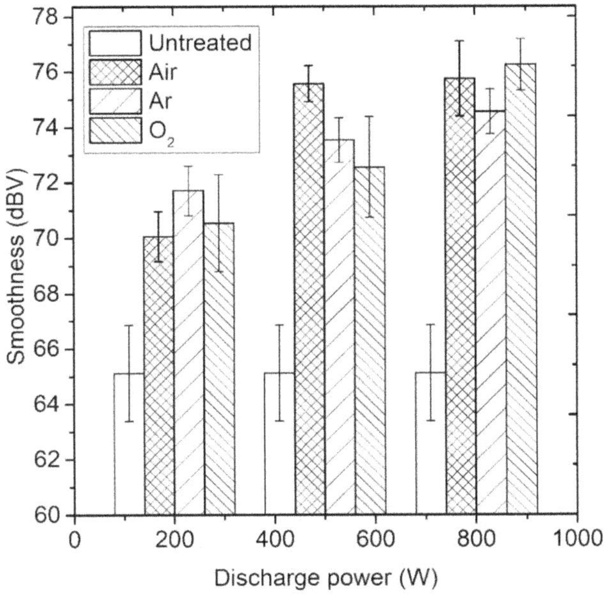


**Figure 4.5.** Effect of plasma treatment on the smoothness property of P/C fabric.

The smoothness of the fabric reduced after plasma treatment due to the formation of voids, cracks and pores on the plasma treated P/C blend fabric surface by the etching and cleaning effect [188]. This gives an uneven fabric surface and comparatively high surface roughness. Prolonged treatment time and high plasma power produced rougher fabric surfaces. As shown in Figure 4.5, all gases used have almost similar influence on the increase in surface roughness as a function of the discharge power. There is no significant difference between the gases either their similar influence or the accuracy of measurements.

### Softness

Softness is a flexural property of the fabric and it is one of the key elements of the clothing comfort [189]. A lower softness value indicates a softer fabric surface. The untreated fabric had a softness value of 10.293 dBV. All plasma-treated samples have a higher value than untreated blend fabric, as shown in Figure 4.6.

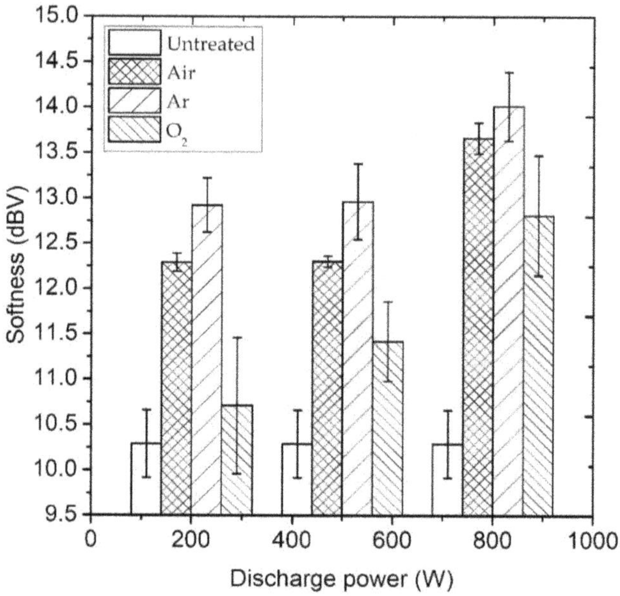


**Figure 4.6.** Influence of plasma treatment on the smoothness property of the fabric.

After plasma treatment, softness decreased due to the etching and sputtering effect of the fiber surface. Additionally, the softness value of the fabric further increases when the plasma power and the exposure time are enhanced, which means that the softness of the blend fabric decreases. As a result, the fabric comfort property slightly affected after plasma treatment [188]. In particular, Ar-plasma treated samples more affected when compared with air and $O_2$-plasma treated samples. Because plasma processing with novel gas can produce harder and stronger substrate surfaces by means of multiple chemical links between the molecular chains of polymers [29].

**Stiffness**

The stiffness values of plasma-treated and untreated P/C blend fabrics are presented in Figure 4.7. According to TSA measurement, the lower the value of stiffness, the stiffer the sample is. Thus, it is a direct indicator of compliance. Stiffness refers to the tendency of a fabric to keep standing without any support or resistance to be bent when external forces applied on it [190]. The untreated fabric had a stiffness value of 1.33 mm/N.

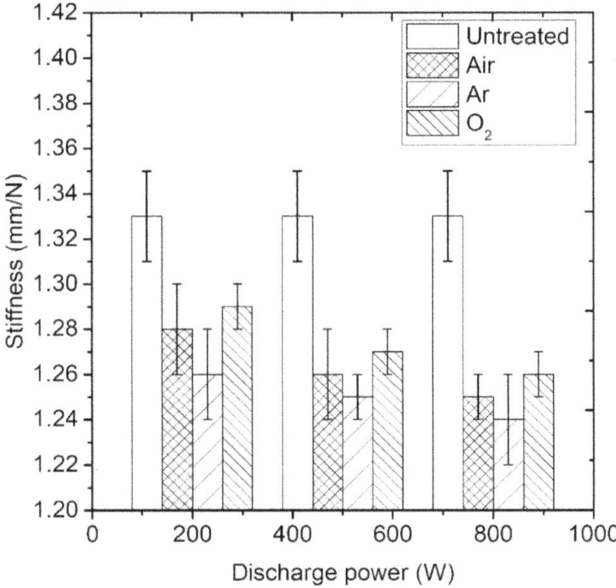


**Figure 4.7**. Influence of plasma treatment on the stiffness property of P/C fabric.

All the plasma-treated samples showed a lower stiffness value than the untreated samples, which demonstrate that plasma treatment causes an increase in stiffness of P/C blend fabrics. In addition, the fabric became stiffer when the plasma variables, power and treatment time, increased. During plasma bombardment, the PET fiber surface etched away when the power and treatment increased; and it became harder when exposed to air. As a consequence, the fiber and the yarn became stiffer and the stiffness of the P/C blend fabric enhanced more intensely. The small detected increase in stiffness after plasma treatment means that the fabric becomes presumably less comfortable to a sensitive person [191].

**Hand-feel (HF)**

The HF value is the overall effect of stiffness, smoothness, softness, number of plies, thickness and weight of fabric, i.e. HF = f (TS7, TS750, D, weight, thickness, number of plies). A higher HF value obtained when the TS7 and TS750 values decreased and D value increased, correspondingly. Additionally, the elasticity, plasticity and hysteresis properties of the fabric were also included in HF values through Emtec software within the TSA device. As is presented in Figure 4.8, all plasma-treated samples had lower HF values than untreated samples.

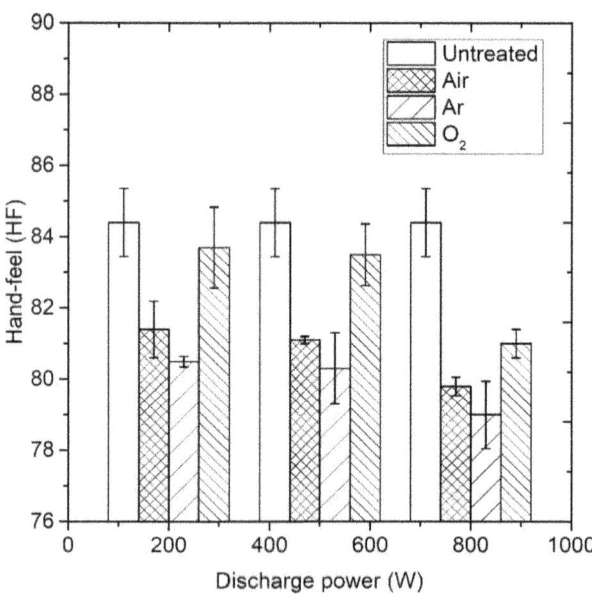


**Figure 4.8.** Effect of plasma treatment on the hand-feel property of the P/C fabric.

The comfort property of the fabric is an undesirable affected. In general, the plasma treatment affected the haptic property of the P/C blend fabric [188, 192]. The degree of its effect depends on the types of plasma gas, duration of treatment and the amount of plasma power.

### 4.2.2 Surface and volume resistivity behavior of the fabric

The surface resistivity ($\rho_s$) is the resistance to leakage current along the surface of an insulating material, while the volume resistivity ($\rho_v$) is the resistance to the leakage current through the whole body of an insulating material under standard conditions. The electrical resistivity of any dielectric material mainly depends on the applied voltage level, relative humidity, temperature and electrification time. Insulation resistivity is a function of both the $\rho_s$ and $\rho_v$ of the substrate [12]. The $\rho_s$ changes almost instantaneously with the change in relative humidity, while $\rho_v$ is particularly sensitive to temperature changes. In this study, both plasma-treated and untreated P/C blend fabrics were conditioned at 65 ± 2% RH and 21 ± 1 °C for 24 h prior to resistivity measurement. The $\rho_s$ and $\rho_v$ measurement were carried out under above-mentioned standard conditions for 60 sec electrification time by applying 100 V on the backside of the blend fabric. This fabric side was directed towards the wearer's body.

P/C blend fabric with plasma treatment would develop low electrical resistivity. During the plasma process, a mixture of electrons, free radicals, excited species, photons and ions can interact, either chemically or physically, with the P/C blend fabric surface. Consequently, the surface may incorporate new chemical groups and/or etch away the surface materials. In this experiment, several factors that remained constant including: set pressure of 0.3 mbar, gas flow rate of 60 sccm and at room temperature. The selection of these parameters was based on previous experience in order to obtain the optimum result [193].

**Effect of the discharge power**

Figure 4.9 shows the effect of discharge power on the $\rho_s$ and $\rho_v$ of the blend fabric after 10 min plasma treatment, respectively. The discharge power was varied from

200 to 800 W during plasma treatment. The mean of three measurements was taken for each datum in the figure. The lower the ρs and ρv is lower the static charge of the sample. However, a test result showed that the $\rho_s$ and $\rho_v$ of untreated P/C blend fabrics developed 14.04 GΩ /sq. and 18.01 GΩ.m, respectively, which inferred that it would adversely affect the wearer's body comfort when using the fabric as a garment. Increased discharge power with the decreased trend of resistivity was observed in Figure 4.9. Different active species from the air, Ar and $O_2$-plasma bombarded and/or reacted with the blend fabric surface by activation, etching, sputtering and cleaning to change its surface properties. As a result of the interaction, carboxyl and hydroxyl groups may be formed. These hydrophilic functional groups would promote the electrical conductivity of the blend fabric. During the treatment process, some low molecular organic materials may be removed by plasma, resulting in an etched feature. The rougher surface would offer space for water molecules, which is conducive to dissipate the build-up static charge in the blend fabric [194].

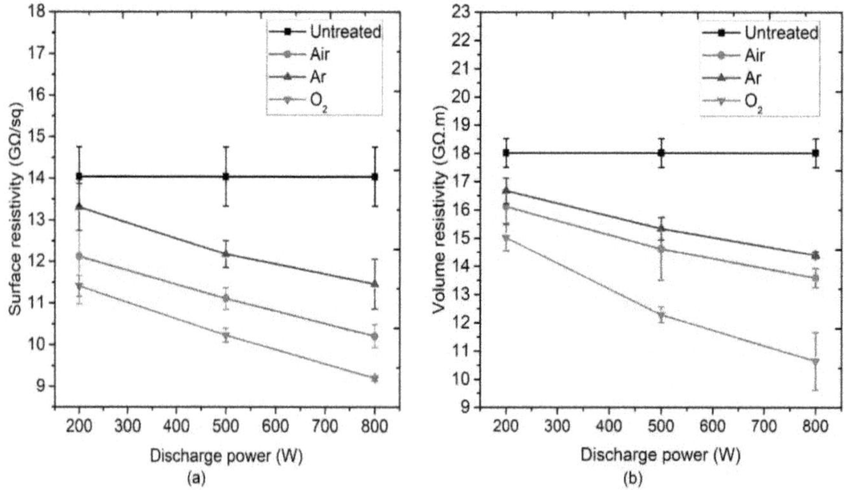


**Figure 4.9.** Effect of discharge power: (a) surface resistivity and (b) volume resistivity of the fabric.

**Effect of treatment time**

Figure 4.10 illustrates the effect of treatment time on the surface and volume resistivity of the blend fabric at 500 W of discharge power. The plasma exposure time varied from 5 to 15 min in the experiment. The data point in the figure was the average of the three readings. It is noticeable that the less electrical resistivity, the better electrical conductivity of the fabric. This implied that the hazard potential of the blend fabric would be lowered. The wearer's body would also be more comfortable. Surface and volume resistivity decreased as the treatment time increased. This is due to the formation of polar groups by air and $O_2$-plasmas, as well as physical changes by argon, air and $O_2$-plasmas on the surface of blend fibers, which would affect the electrical resistivity. These effects would promote the adsorption of moisture on the fiber surface and reduce its resistivity [195].

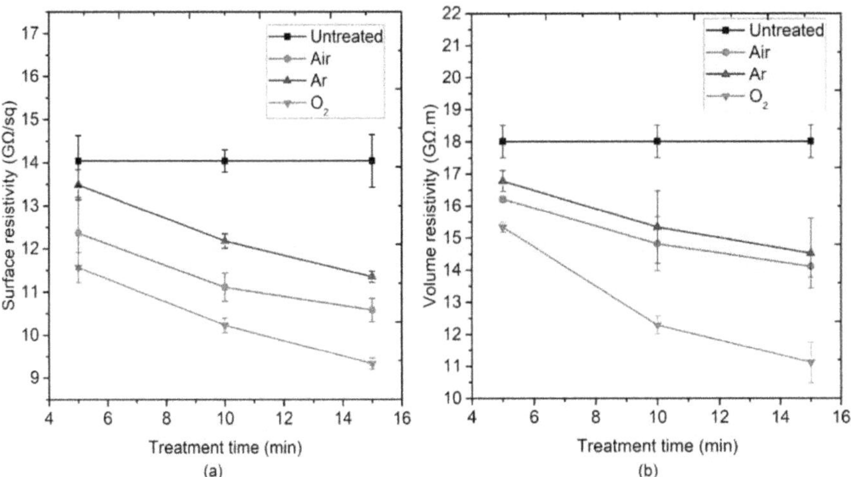


***Figure 4.10.*** *Influence of treatment time: (a) surface resistivity and (b) volume resistivity of the fabric.*

**Effect of the types of plasma gases**

In order to study the effect of gas type on the electro-physical properties of the plasma-treated P/C blend fabric, a series of treatments with Ar, air and $O_2$ gases were carried out keeping other parameters constant. The plasma-modification was performed in various discharge powers (200, 500 & 800 W) and the results of $\rho_s$ and $\rho_v$ of the blend fabric, after 10 min plasma treatment, are shown in Figure 4.9 (a) and (b), respectively. Under a discharge power of 800 W, $O_2$-plasma treated samples developed a $\rho_s$ value of 9.2 GΩ/sq., compared to the 14.04 GΩ/sq. of the untreated samples with a reduction of 35.5%. In addition, the $\rho_s$ decreased significantly to 10.21 GΩ/sq. for air-plasma and 11.46 GΩ/sq. for Ar plasma-treated samples, resulting in a 27.3% and 18.4% reduction, accordingly, as depicted in Figure 4.9 (a). The $\rho_v$ of $O_2$-plasma treated sample at 800 W highly reduced to 10.65 GΩ.m, compared to the 18.01 GΩ.m produced by untreated samples with a reduction of 40.9%. Air and Ar plasma treated samples also generated a $\rho_v$ of 14.35 and 14.4 GΩ.m. These values correspond to a 20.3% and 20% reduction, respectively, as presented in Figure 4.9(b).

Plasma treatments were also carried out in various exposure times (5,10 &15 min) and the results of $\rho_s$ and $\rho_v$ of the blend fabric, using 500 W discharging power, are shown in Figure 4.10 (a) and (b), respectively. A similar effect is observed, like that of discharge power, by means of Ar, air and $O_2$ plasmas on electrical properties of P/C blend fabric. For instance, 15 min Ar, air and $O_2$-plasmas treatment gave 11.35, 10.58 and 9.33 GΩ/sq. of $\rho_s$ as well as 14.52, 14.11 and 11.12 GΩ.m of $\rho_v$ compared to 14.04 GΩ/sq. and 18.01 GΩ.m of the untreated P/C blend fabrics, respectively.

In general, the changes in the fabric surface caused by plasma are predominantly depending on the gas type and conditions of plasma generation [196]. As discussed above, the influence of $O_2$ plasma on the resistivity of blend fabric is the highest, whereas the effect of Ar plasma is the least. It is correlated to the nature of gases. In particular, $O_2$ plasma is oxidizing in nature and introducing oxygen functional groups, including C−O, C=O, O−C−O and C−O−O at the surface. This leads to the surface to be hydrophilic and reduces the resistivity of the fabric. In $O_2$ plasma, etching of the fiber surface and formation of oxygen functional groups at the fiber surface occur simultaneously [53]. Air plasma also gave a similar result like $O_2$ plasma. This is due to 21% of $O_2$ and 78% of $N_2$ composition in the air. Nitrogen-containing functional groups also may enhance the wettability of the fabric like $O_2$ plasma [197]. Contrast, Ar plasma can perform surface cleaning by etching and sputtering processes. In addition, Ar plasma treatment, at low power and for certain periods, sufficiently remove $H_2$ and form free radicals near or at the surfaces which then interact to form cross-links and unsaturated groups with chain scission, but it cannot directly incorporate functional groups [43, 197–199]. That is why Ar plasma had the least effect on the resistivity of the P/C blend fabric.

## 4.3 Wettability and aging behavior of plasma treated fabric

### 4.3.1 Effect of aging on wettability of Ar, air and $O_2$-plasma treated fabric

The effect of aging on wettability of Ar, air and $O_2$-plasma plasma treated and untreated P/C blend fabric was performed by TEGWA drop test method. All treated, at the condition of 500 W, 0.3 mbar, 10 min and 60 sccm, and untreated samples were stored under standard condition for 1, 7 and 14 day/s. The results of drop solution sink-in time and spread area are given in Figure 4.11. The dye solution on untreated fabric sunk-in within 28.6 sec, whereas for all treated samples, it was absorbed and disappeared in less than 10 seconds. After a 14 days aging period, the absorption time of plasma treated samples increased from 5.36 to 7.75 sec for oxygen, 5.74 to 8.92 sec for air and 6.36 to 9.72 sec for argon plasma treatments as shown in Figure 4.11(a). Upon aging, the surface polarity is reversed and hydrophic surfaces are recovered due to the mobility of volatile low molecular weight hydrolyzed fragments, the reaction of residual radicals continuously with the ambient environment during storage and the decreasing of surface free energy through the passage of time [200]. The rate of aging decreased gradually, but plasma treated surface have better wettability than untreated surface even after long period of time [201].

On the other hand, the spread area of the completely dried color spot on untreated fabric was 1240 $mm^2$. However, the spread area of plasma treated samples decreased from 1741 to 1570 $mm^2$ for oxygen, 1684 to 1543 $mm^2$ for air and 1610 to 1500 $mm^2$ for argon plasma treatments as depicted in Figure 4.11(b). Although all treated samples had a much better hydrophilic behavior, oxygen plasma was more effective compared to air and argon plasma treatments. This can be explained by the formation of more polar groups for instance, hydroxyl and carboxylic on the blend fabric surfaces [202].

As is shown clearly in Figure 4.11, the rate of aging sharply increased in the first 7 days, whereas it slowly decreased for the last 7 days. This indicates that the stability of active species generated on the blend fabric significantly depends on the aging time [203]. Nevertheless, the wettability of plasma treated fabric was better than untreated fabric even after two weeks. Similarly, after 14 days, the spread area of plasma treated samples were smaller compared to freshly treated fabrics, which revealed that the hydrophilicity of the treated fabrics reduced after longer storage times. When the treated P/C blend fabrics were placed in standard condition for some days, the surface energy decreased due to the reorientation of the polar groups in the bulk matter and migration of low molecular weight segments to the surface [202].

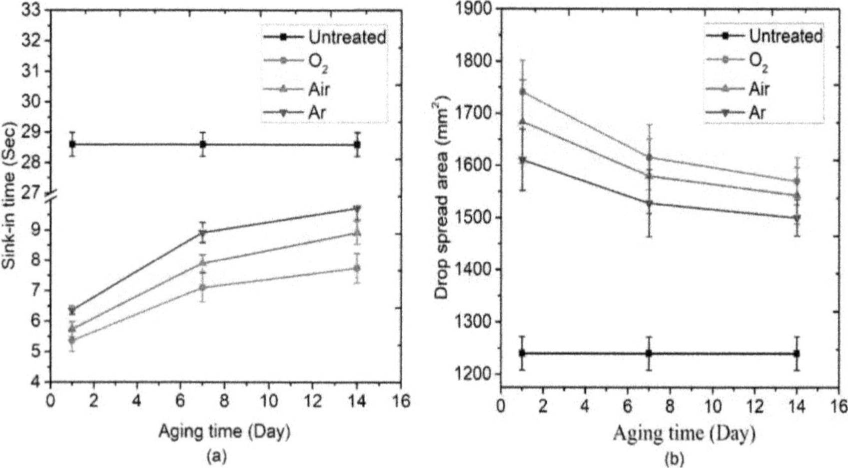


**Figure 4.11.** Evaluation of aging effect on the plasma treated fabric by (a) sink-in time and (b) drop spread area.

Figure 4.12 shows the photographs of dropped and dried color spot appearances on untreated, Ar-, air-, and $O_2$-plasma treated P/C blend fabric samples after 24 h conditioned. As depicted in Figure 4.12, the dye solution spread and absorbed into the fabrics, but the degrees of hydrophilicity of the samples were not the same.

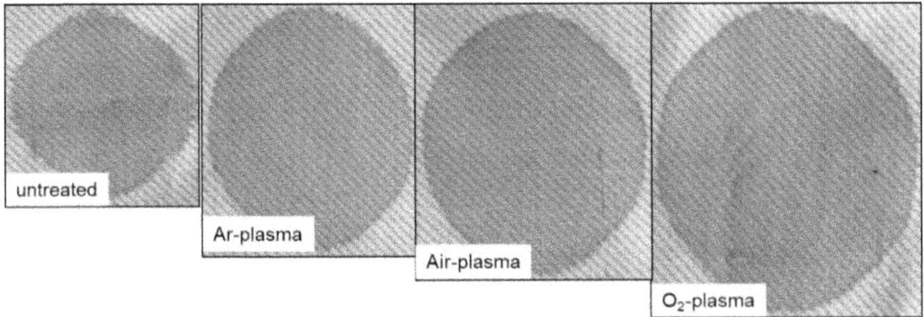


**Figure 4.12.** Illustration of drop spot appearance photo (1.12:1) on the fabric before and after plasma.

As it is seen in Figure 4.11 (quantitatively in the form of sink-in time and drop spread area) and Figure 4.12 (qualitatively), the difference is clear and visible. In particular, the border of the color spot on the untreated sample was unevenly serrated even if it was slightly round shape characteristics while, Ar-, air-, and $O_2$-plasma treated samples had an oval shape due to the better horizontal wicking in the warp direction and did not show serrated features on the border of the color spot except the difference of sink-in time and drop spread area.

### 4.3.2 Wettability of plasma coated P/C blend fabric

Plasma-enhanced chemical vapor deposition of $O_2$/HMDSO, $N_2$/HMDSO, $H_2O$ /HMDSO and HMDSO were carried out on P/C blend fabrics. Firstly, the surface was cleaned and activated for 5 min by applying oxygen (50 sccm), nitrogen (50 sccm) and water vapor (liquid flow of 300 µL/min) plasmas. Then, the plasma coatings were performed at the condition of 700 W, 10 min and 0.3 mbar for $O_2$/HMDSO, $N_2$/HMDSO and $H_2O$/HMDSO plasmas, whereas HMDSO plasma treatment was taken 15 min without surface cleaning and activation. In both cases, the pulse of vaporizers performed at 10 seconds of cycle time and 25% of due time.

Figure 4.13 shows the influence of plasma coating on the wettability of P/C blend fabrics. All of the plasma irradiated samples exhibited much higher drop water sink-in time compared to the untreated substrates except $H_2O$/HMDSO plasma coated samples, measured by TEGWA drop test method, as depicted in Figure 4.13(a). Basically, HMDSO is mostly used to impart water repellent finish to the textile substrate. As a result, the dye solution had taken more than 923 seconds to be sunk-in on HMDSO coated fabric. This is due to the orientation of Si-group and methyl groups on the fabric [204]. In addition, $N_2$/HMDSO, $O_2$/HMDSO and untreated fabric had 415.87, 371 and 28.6 s sink-in time, respectively. However, surprisingly, $H_2O$/HMDSO plasma coated fabric had very short sink-in time (i.e. 14.46 s) compared to untreated and other treated samples, as shown in 4.13(a). Most likely, this is a result of a higher ratio of water vapor to HMDSO compared to the ratio of oxygen and nitrogen to HMDSO.

Similarly, the drop spread area of plasma coated samples are also presented in Figure 4.13(b). The area of completely dried color spot on $H_2O$/HMDSO plasma coated fabric was 1656 $mm^2$, while the drop spread area of HMDSO plasma coated fabric was 1158.67 $mm^2$. This indicates a significant difference between the wettability of the two plasma coated fabrics. The spread area of $N_2$/HMDSO and $O_2$/HMDSO plasma treated fabric showed more or less similar results, but untreated fabric had 1240 $mm^2$. Among all plasma coated fabrics, $H_2O$/HMDSO treated samples had a much better hydrophilic behavior. This is the influence of water vapor plasma on chemical and morphological properties of P/C blend fabric. In particular, C–O, O–C–O, C=O, and O=C–O bonds can be formed on the fabric surface [205].

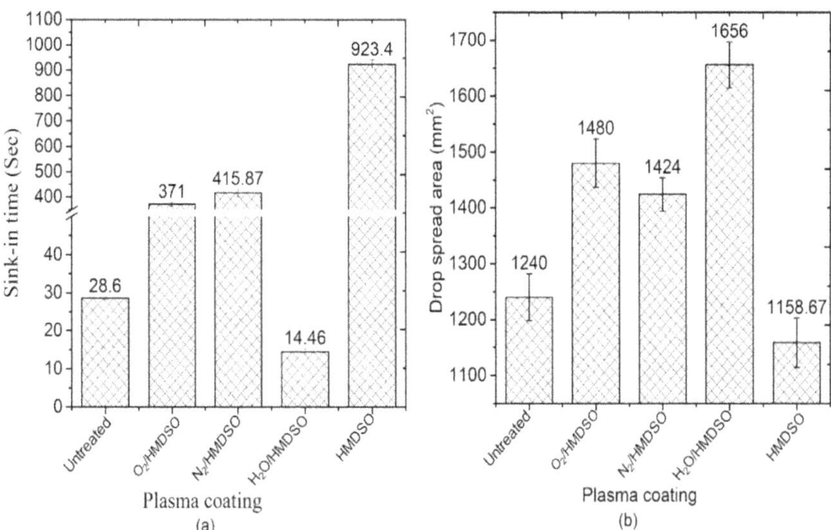


**Figure 4.13**. Wettability behavior of plasma coated fabric: (a) sink-in time and (b) drop spread area.

The photographs of dropped color spot appearances on HMDSO, $N_2$/HMDSO, $O_2$/HMDSO, $H_2O$/HMDSO plasma coated and uncoated P/C blend fabrics, after 24 h conditioned, are displayed in Figure 4.14. In general, a hydrophilic treated fabric exhibits round or oval run characteristics. The drop falls positively into the fabric, and first of all produces, as illustrated Figure 4.14, a spread of different diameters. During air drying, the liquid migrates further, but the extent of spread measured after the drop has sunk in is also visible on the dry test sample in terms of a higher color intensity. $H_2O$/HMDSO plasma coated fabric showed uniform flow behavior of dye solution and produced more evenly and deeper color shade compared to others. However, HMDSO plasma coated fabric had a more unevenly serrated border of color spot and very light shade. It is the hydrophobic nature of HMDSO monomers. In addition, an unevenly serrated border of color is visible in uncoated, $N_2$/HMDSO and $O_2$/HMDSO plasma coated fabrics at different levels. When the sink-in time increases, the drop spread area decreases. It is true for non-reactive plasma gases. However, in the typical case of used plasma polymerizing gas, this may be often untrue. The reason is the change of surface morphology due to the deposition of plasma polymer on the surfaces. Consequently, the porosity of the fabric surface decreased and some channels were formed over it along the fiber axes on $O_2$/HMDSO and $N_2$/HMDSO plasma polymerized surfaces. This channel may assist the horizontal wicking and increased the drop spread areas.

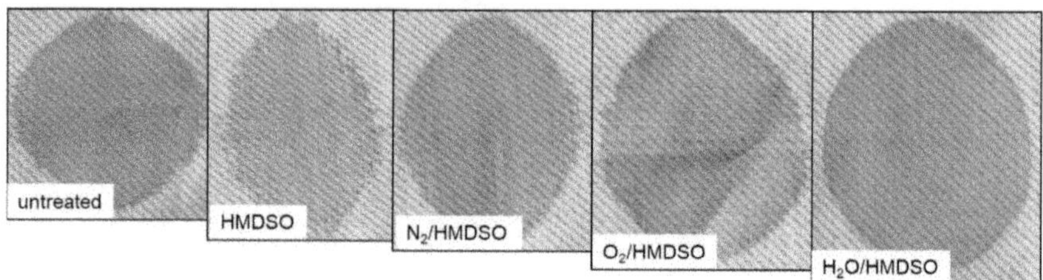


**Figure 4.14.** Photographs (1.34:1) of drop spot appearance on plasma coated and untreated fabrics.

## 4.4 Tensile strength of P/C blend fabric

The influence of plasma application on the tensile characteristics of the P/C blend fabric was evaluated by determination of maximum force and elongation at maximum force using the strip method. The results of tensile strength and elongation of the blend fabrics are shown in Figure 4.15. The tensile strength and elongation of the plasma-treated P/C samples are comparatively lower than the untreated samples in the warp direction, whereas in the weft direction, there is no significant difference between plasma treated and untreated fabrics as presented in Table 4.6.

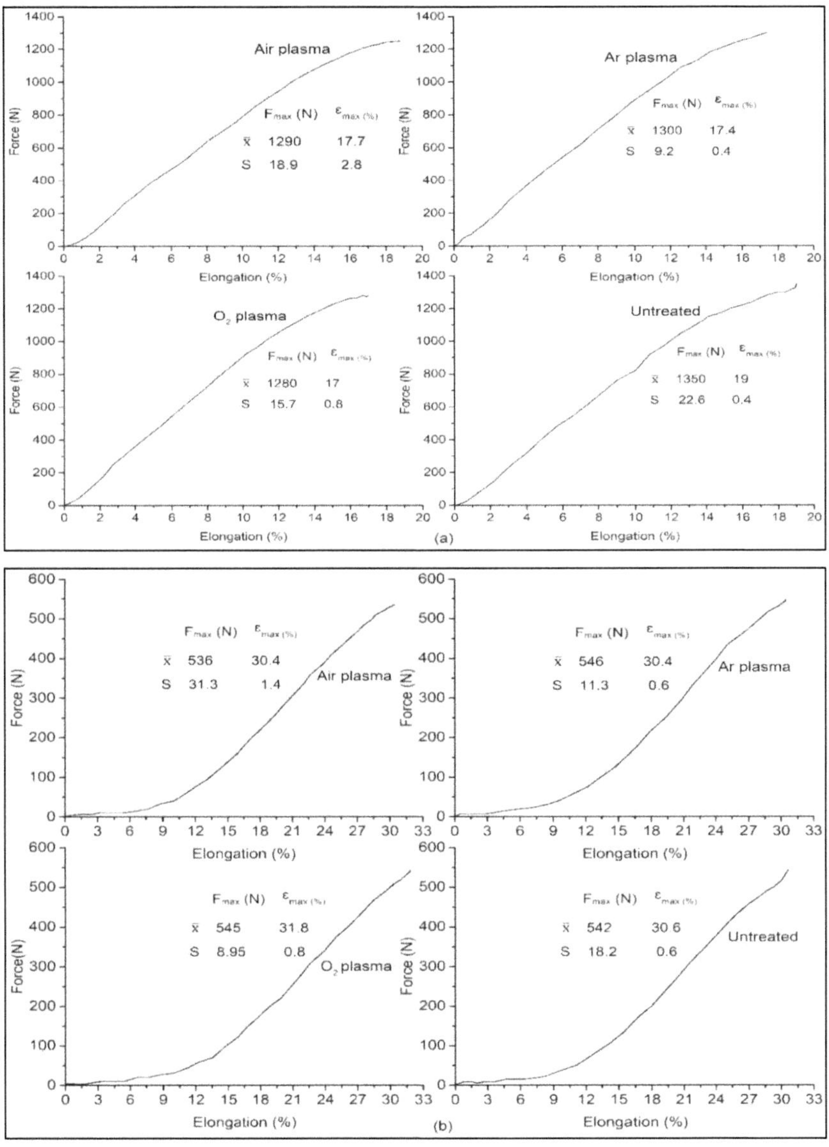


**Figure 4.15.** Tensile strength and elongation of air, Ar, $O_2$ plasmas treated and untreated blend fabric in the warp (a) and weft (b) directions.

The results of One-way ANOVA for tensile strength and elongation of air, Ar, $O_2$ plasmas treated and untreated P/C blend fabrics are summarized in Table 4.6. It can be clearly seen that in warp direction, the mean values for tensile strength and elongation (Table 4.6. [a] and [c]) are significantly ($p < 0.05$) different from each other for the 95% confidence level. In contrast, the tensile strength and elongation values for the weft direction (Table 4.6. [b] and [d]) are not significantly different from each other ($p < 0.05$) for the 95% confidence level.

**Table 4.6:** One-way ANOVA on tensile strength and elongation of air, Ar, $O_2$ plasmas treated and untreated P/C blend fabric.

| Source of variation | DF | Sum of Squares | Mean Square | F-Value | Prob>F |
|---|---|---|---|---|---|
| Between groups | 3 | 14920 | 4973.33 | 28.83 | 1.0927E-8[a] |
| Within groups | 16 | 2760 | 172.5 | | |
| Total | 19 | 17680 | | | |
| Between groups | 3 | 322.55 | 107.52 | 0.29 | 0.8348[b] |
| Within groups | 16 | 6014 | 375.88 | | |
| Total | 19 | 6336.55 | | | |
| Between groups | 3 | 10.742 | 3.58 | 15.70 | 5.0119E-5[c] |
| Within groups | 16 | 3.65 | 0.23 | | |
| Total | 19 | 14.39 | | | |
| Between groups | 3 | 7.53 | 2.51 | 3.18 | 0.0525[d] |
| Within groups | 16 | 12.62 | 0.79 | | |
| Total | 19 | 20.15 | | | |

[a]Tensile strength in warp direction; [b]tensile strength in weft direction; [c]elongation in warp direction; [d]elongation in weft direction.

The effect of plasma treatment on the tensile characteristics of the P/C blend fabric was also evaluated by determination of maximum force and elongation at maximum force using the strip method for the mixture of gases with HMDSO as shown in Figure 4.16. The tensile strength and elongation of the plasma-treated P/C samples are relatively smaller than the untreated samples in the warp direction, whereas in the weft direction, there is no significant difference between plasma treated and untreated fabrics as presented in Table 4.7.

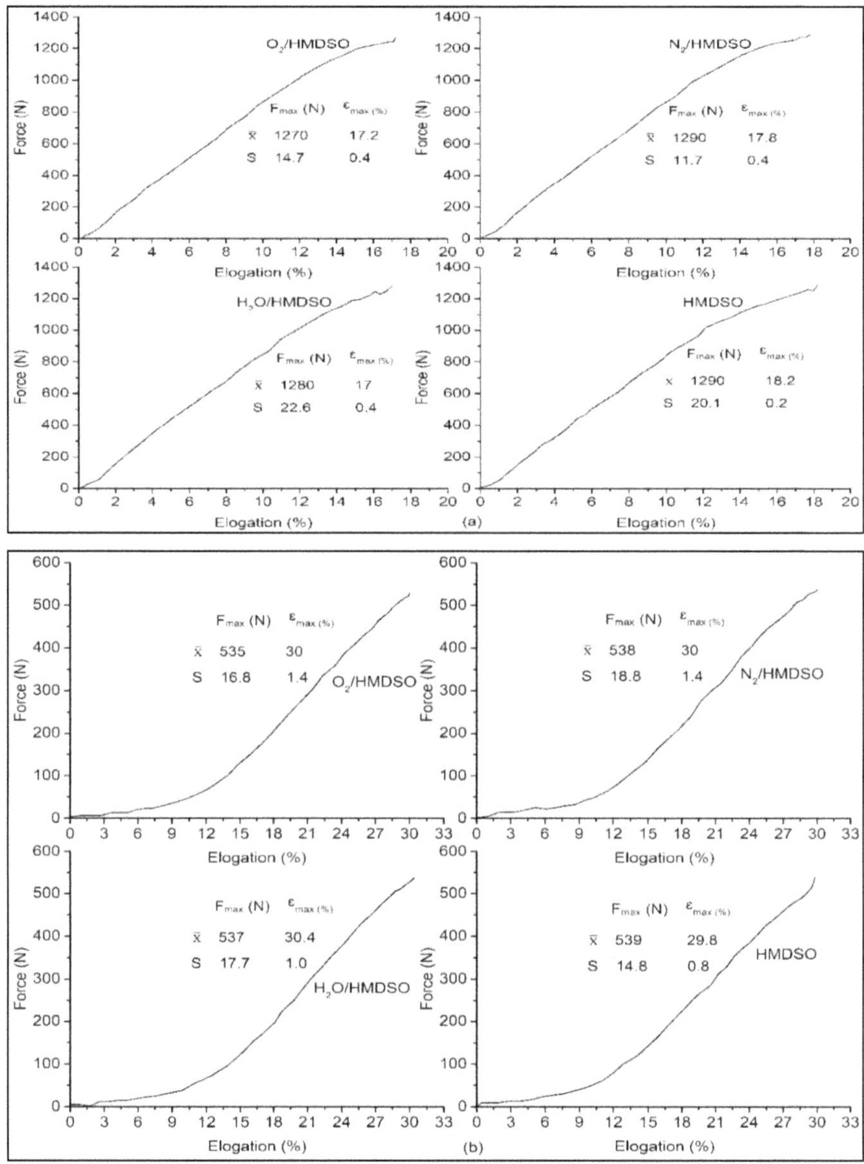


**Figure 4.16.** Tensile strength and elongation of $O_2$/HMDSO, $N_2$/HMDSO, $H_2O$/HMDSO and HMDSO plasma coated P/C fabric in (a) the warp and (b) weft directions.

The results of One-way ANOVA for tensile strength and elongation of $O_2$/HMDSO, $N_2$/HMDSO, $H_2O$/HMDSO, HMDSO plasmas treated and untreated P/C blend fabrics are summarized in Table 4.7. It clearly shows that the means of the experimental result in the weft direction for tensile strength and elongation are not significantly different ($p<0.05$) at 95% confidence level. However, the effect of plasma coating on tensile strength and elongation in the warp direction are significantly different ($p<0.05$) at 95% confidence level. In both plasma treatments i.e. using gases and mixtures of gases with HMDSO monomer, the significant differences in warp direction came from untreated fabric samples. The crimp % of warp yarns are lower than weft yarns. Probably, this may contribute more surface areas of the warp yarns are exposed to plasma treatment, which leads to the part of the fiber surface to be damaged. Consequently, the breaking strength and elongation of blend fabric in warp direction were decreased compared to untreated fabric.

Although the tensile strength and elongation of the fabric is statistically significant in the warp direction, it may not necessarily be the effect of plasma treatment. Since the plasma application is a surface treatment and shouldn't affect the bulk properties of the fabric significantly as confirmed in the weft direction [205, 206].

**Table 4.7:** One-way ANOVA on tensile strength and elongation of $O_2$/HMDSO, $N_2$/HMDSO, $H_2O$/HMDSO, HMDSO plasmas coated and untreated P/C fabric.

| Source of variation | DF | Sum of Squares | Mean Square | F-Value | Prob>F |
|---|---|---|---|---|---|
| Between groups | 4 | 19704 | 4926 | 17.22 | 2.9076E-6[a] |
| Within groups | 20 | 5720 | 286 | | |
| Total | 24 | 2542 | | | |
| Between groups | 4 | 141.36 | 35.34 | 0.10 | 0.9802[b] |
| Within groups | 20 | 6871.6 | 343.58 | | |
| Total | 24 | 7012.96 | | | |
| Between groups | 4 | 14.20 | 3.55 | 28.27 | 5.5858E-8[c] |
| Within groups | 20 | 2.51 | 0.13 | | |
| Total | 24 | 16.71 | | | |
| Between groups | 4 | 2.1 | 0.52 | 0.46 | 0.7642[d] |
| Within groups | 20 | 22.78 | 1.14 | | |
| Total | 24 | 24.88 | | | |

[a]Tensile strength in warp direction; [b]tensile strength in weft direction; [c]elongation in warp direction; [d]elongation in weft direction.

Sometimes, this kind of results may be obtained due to the difference of the number of yarn threads within the prepared samples. It significantly influenced the tensile strength and elongation of the fabric more than the effect of plasma treatment on the bulk properties. Since the fabric strength and elongation are considerably affected by yarn property and fabric construction, which remain unaltered by plasma treatment [207]. In order to solve this type of limitation, the individual yarns can be unraveled from the fabric to measure the tensile strength rather than testing the fabric strips [206].

## 4.5 Surface characterization of P/C blend fabric

### 4.5.1 Surface morphology analysis

SEM is the well-known and most widely used tool for surface morphological analyzes. The SEM images of (a) untreated, (b) Ar, (c) $O_2$ and (d) air plasmas treated P/C blend fabric samples are shown in Figure 4.17. The single fiber was scanned at the scale of 1 µm and a magnification of 10,000X with 5 kV to present the surface characteristics. It can be seen on SEM images that untreated PET and cotton fibers have relatively flat and smooth surfaces, whereas after 10 min plasma treatment, the fibers surface have become full of new features according to the types of plasma gases [208]. From Figure 4.17, nano-cracks and voids are found on the treated fibers surface due to plasma etching [209].

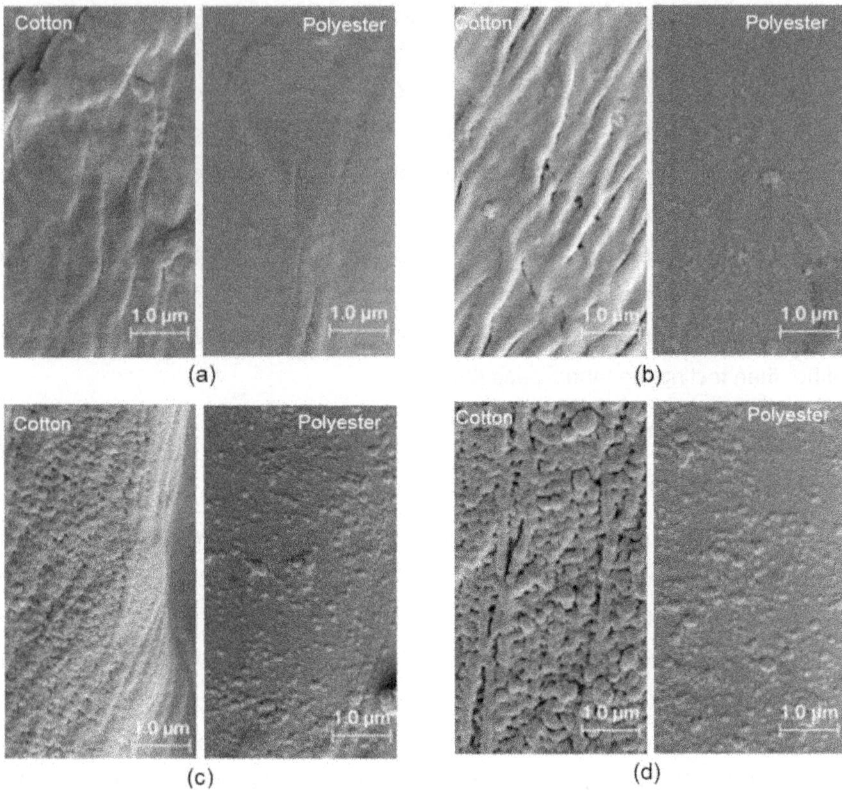


**Figure 4.17.** SEM images: (a) untreated, (b) Ar, (c) $O_2$ and (d) air plasma treated P/C fabric at LPP conditions (500 W, 0.3 mbar and 10 min).

The morphological changes of the fiber surfaces were produced by ion bombardment with high energetic particles during the plasma process. In particular, air-plasma (Figure 4.17d) created porosity and voids on cotton fibers, as well as solidified nano-spots on the PET fiber surfaces [73]. However, Ar-plasma (Figure 4.17b) produced grooves and ribbon-like features on the cotton surface and eroded surfaces on the PET fibers [29, 210]. The untreated polyester fiber had a smooth surface while the surface of all plasma-treated fibers became rougher. Nodule like features appeared on the plasma-treated fiber surface. These nodules were distributed densely specially on the surface of oxygen and air plasma-treated polyester fiber [211]. The polyester fiber treated with plasma had an evident change with the presence of nodules like feature in the fiber surface morphology. The roughness indicated that etching action had occurred to the fiber surface. Therefore, the rough surface could provide more capacities for the blend fabric to capture water and also facilitate the penetration of water into the P/C blend fabric [212]. In addition, $O_2$-plasma (Figure 4.17c) also developed a similar feature to that of the air-plasma treatment. The effect of air-plasma treatment is more pronounced compared to $O_2$-plasma [213]. Ar-plasma produced stronger and rougher surfaces due to the cross-linking polymers. In all plasma treatments, some solidified nano-spots are observed on the plasma-treated PET fiber surfaces [194]. These confirmed that plasma treatments have a potential to change the surface of P/C blend fabric.

The SEM was also used to investigate the degree of deposition of thin films on the surface of P/C blend fabric. Figure 4.18 shows typical SEM images of purely HMDSO coated and the mixture of HMDSO monomers with oxygen, nitrogen and water vapor plasma coated polyester and cotton fibers. It can be also seen from Figure 4.18 (e) that the topography of untreated cotton and polyester fibers are without any sign of polymer deposition. SEM images of all plasma coated P/C samples i.e. $O_2$/HMDSO (4.18 (a)), $N_2$/HMDSO (4.18 (b)), $H_2O$/HMDSO (4.18 (c)) and HMDSO (4.18 (d)) plasmas coated exhibit the layer of polymeric materials on the fibers surface [206]. Moreover, it appears that the amount of deposition increased with an increase of the mass flow of the gases rather than the monomer. It is directly correlated to the hydrophilic properties of plasma polymerized P/C blend fabric [169]. In this, plasma–enhanced chemical vapor deposition (PECVD), the mass flow of nitrogen, oxygen and water vapor gases were 63, 71 and 300 mg/min, correspondingly, while the mass flow of HMDSO monomer was less than 5 mg/min. That is why the $H_2O$/HMDSO plasma coated sample was more hydrophilic compared to others coated samples.

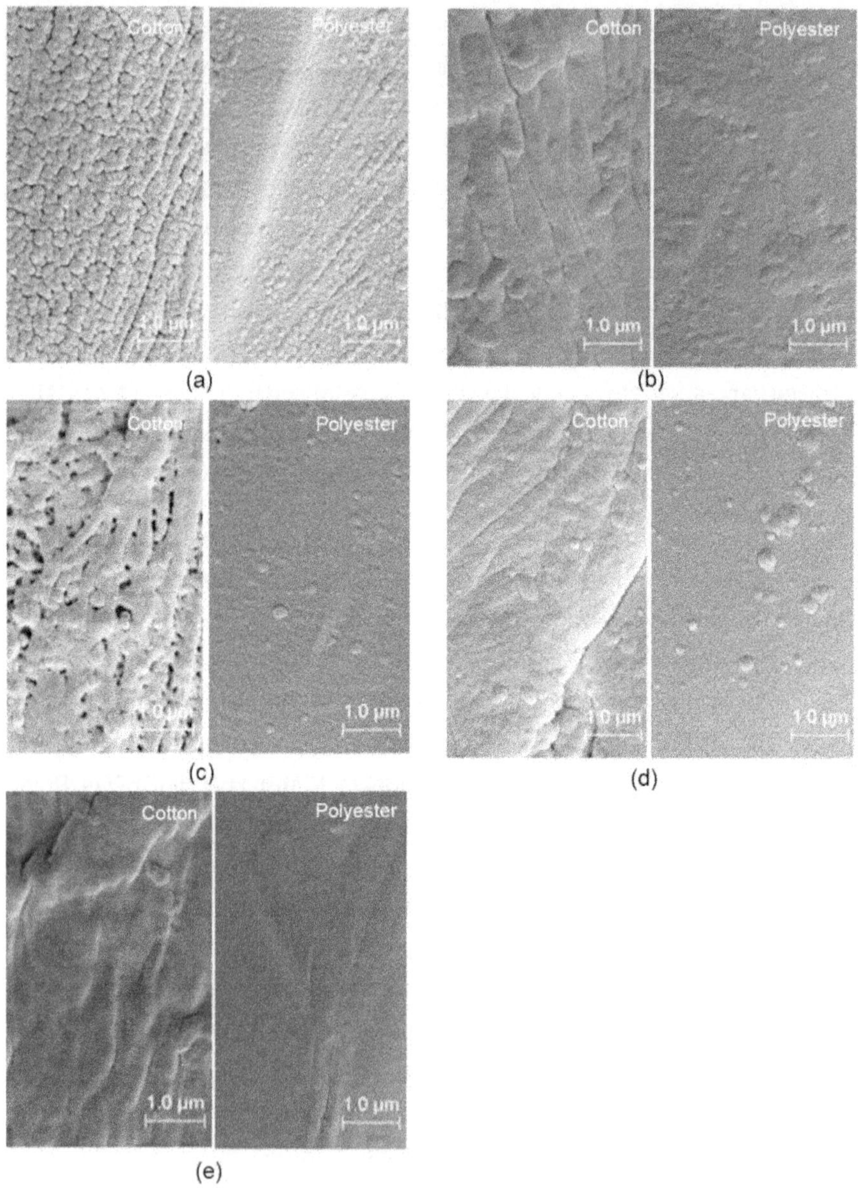


**Figure 4.18.** SEM images: (a) $O_2$/HMDSO, (b) $N_2$/HMDSO, (c) $H_2O$/HMDSO, (d) HMDSO plasmas coated and (e) untreated P/C fabric.

The fibrillar structure of the cotton fibers are partially covered by $O_2$/HMDSO and $H_2O$/HMDSO plasma coating, as shown in 4.18 (a) and 4.18 (c), whereas $N_2$/HMDSO and HMDSO plasma coated samples are fully covered due to the thin layers of the plasma polymer on the surfaces. In all treated polyester fibers, the nano-particles are formed. Consequently, the high amount of deposition of polymer is responsible for conferring hydrophobic character. The difference of wettability of the samples observed during the water drop test appears to be correspondingly with the SEM images [169].

### 4.5.2 Surface chemical analysis

**Energy-dispersive X-ray (EDX) analysis**

SEM-EDX was employed to study the surface morphology and elemental composition of plasma coated and uncoated P/C blend fabrics. In particular, the atomic % of the plasma polymerized and reference fabrics determined by EDX analysis are shown in Figure 4.19. EDX analysis demonstrated that untreated P/C fabric (Figure 4.19 (a)) was composed of about 62.05% C and 37.95% O. $O_2$/HMDSO coated fabric (Figure 4.19 (b)) presented 59.07% C, 40.40% O and less than 1% of Si and Na, whereas in $N_2$/HMDSO plasma coated (Figure 4.19 (c)), 60.62% C, 38.36% O and 1.02% Si were identified. The types of element presented in HMDSO plasma coated fabric (Figure 4.19 (e)) was similar to $O_2$/HMDSO polymerized fabric. The atomic percentages of HMDSO polymerized fabric consisted of about 60.68% C, 38.31% O and less than 1% of Si and Na. $H_2O$/HMDSO coated fabric (Figure 4.19 (d)) also depicted 63.88% C, 36% O and 0.12% of Si. The introduction of Si atoms and Si- containing groups on the fiber surfaces are responsible to the hydrophobic properties of the blend fabric [169], [214–216].

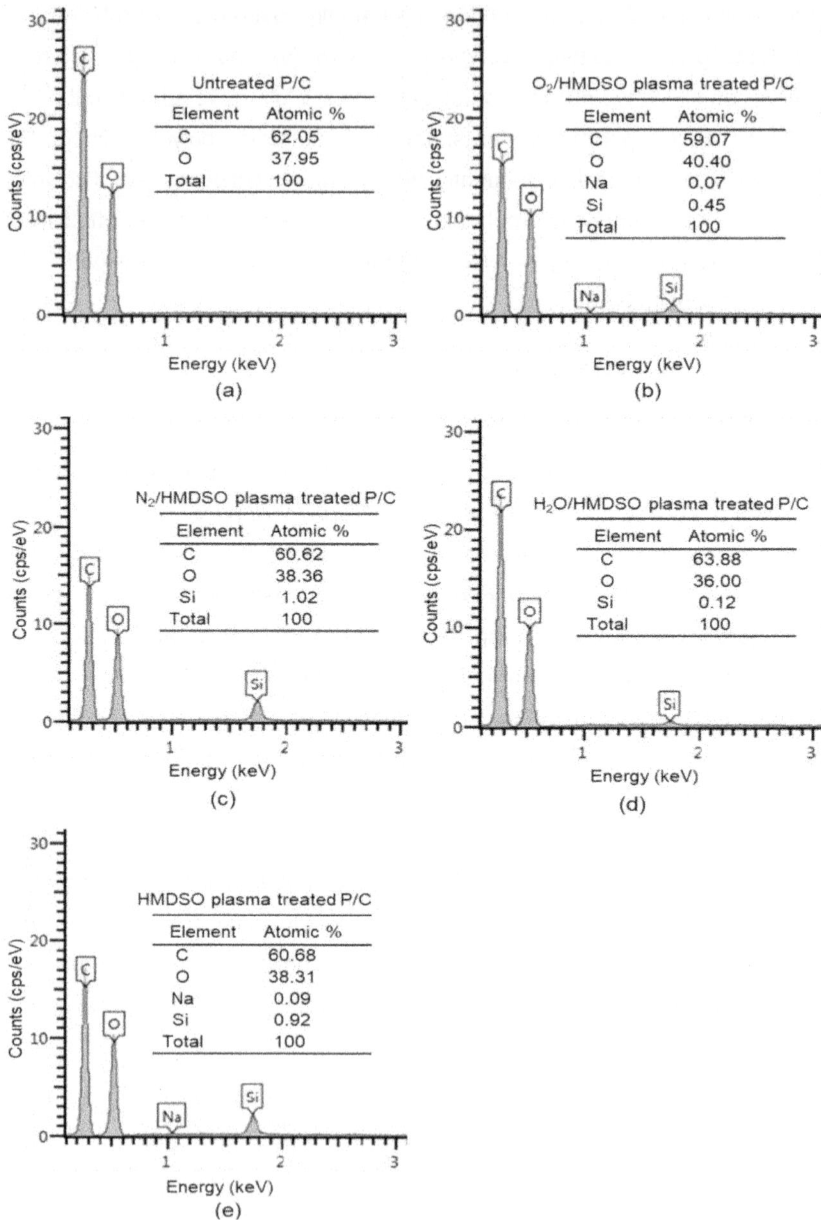


**Figure 4.19.** EDX analysis of P/C blend fabric.

For all plasma coated P/C blend fabric surfaces, Si element was detected in addition to the reference fabric, which confirms that it is introduced into the polymeric film deposited on the fibers surface [217]. Trace amount of Na is also observed in $O_2$/HMDSO and HMDSO plasma coated P/C fabric surfaces [218]. For the untreated and plasma coated samples, only C and O which are the main constituents of the P/C fabrics. The atomic ratio of O/C for untreated, $O_2$/HMDSO, $N_2$/HMDSO, $H_2O$/HMDSO and HMDSO is 0.612, 0.684, 0.633, 0.564 and 0.631, respectively. As expected, the EDX analysis revealed the changed quantitative and qualitative surface composition after deposition of the plasma polymer. Figure 4.19 shows that the O content increased by 1% after plasma coating except $H_2O$/HMDSO. In the case of $H_2O$/HMDSO plasma, O reduced by 1%. However, its hydrophilic property increased as confirmed by drop sink-in time. Therefore, this result may be the limitation of EDX analysis [219].

The EDX analysis is correlated to the interaction of the electron beam with the fabric. The volume where the X-rays are excited and the information about the fabric is gathered depending on fabric properties, such as density, chemical composition and orientation of the sample to the detector. These parameters can influence the count rate. Hence, the exact determination of the analyzed volume is quite difficult. Indeed, the O atomic percentage increased while the C atomic percentage decreased. The increase of the O/C ratio mainly indicates the introduction of oxygen functional groups due to plasma treatment. The high amount of O content led to an improvement in the hydrophilicity of the fabric [220].

**ATR-FTIR analysis**

The ATR-FTIR spectra of air, Ar, $O_2$-plasma treated and untreated P/C fabric samples are presented in Figure 4.20. The blend ratio of the PET fiber in the blend fabric covers 65%. For this reason, the developing peaks of the P/C blend fabric are more or less similar to the peaks of PET. The peaks emerging in the ATR-FTIR spectra at 1712, 1242 and 721 $cm^{-1}$ can be assigned to the chemical nature of the PET fiber [8, 221], whereas peaks observed at 3330 and 2902 $cm^{-1}$ can be associated to both PET and cotton in the blend. Nevertheless, the absorption bands and their functional groups of plasma treated and untreated P/C blend fibers are presented below in Table 4.8 [222].

**Table 4.8:** Functional groups and their peaks found in the ATR-FTIR spectrum of plasma treated and untreated P/C blend fabric.

| **Absorption peaks ($cm^{-1}$)** | **Functional groups** |
|---|---|
| 3330 | Hydroxyl (O-H) |
| 3100-3000 | Aromatic (C-H) |
| 2902 | Aliphatic (C-H) |
| 1712 | Carboxylic (C=O) |
| 1408 | Aromatic ring |
| 1339 | Alkane ($CH_2$) |
| 1242 | Carboxylic (C-O) |
| 1098, 1018 | Ester (O=C-O-C) |
| 872, 721 | Aromatic (C-H) |

The spectra developed from untreated and plasma treated P/C samples are very similar. It does not depict any considerable difference on the surface chemistry of the fabric. However, the intensity ratio was calculated by normalizing the peaks at 1712 and 1242 $cm^{-1}$ with respect to that at 721 $cm^{-1}$.

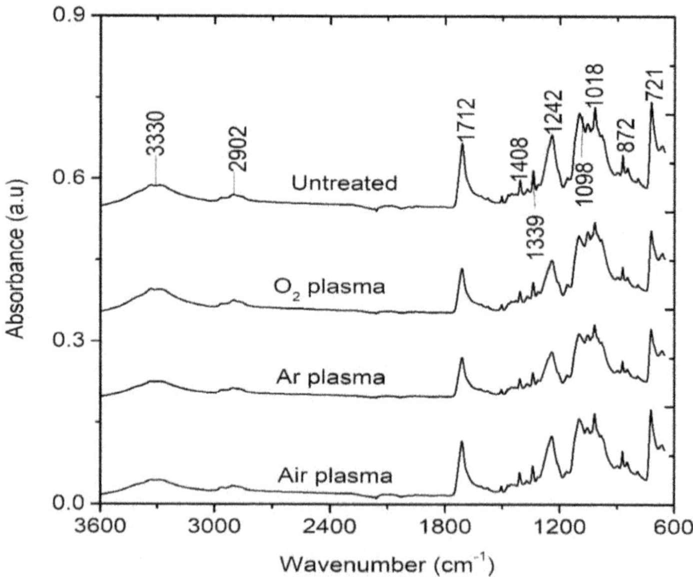


**Figure 4.20.** ATR-FTIR spectra of plasma treated and untreated P/C blend fabric (Plasma condition: 0.3 mbar, 10 min, 500 W and 60 sccm).

The intensity of peak at 721 $cm^{-1}$ was found to be least influenced by various plasma treatments [169]. Thus, it was considered as a reference peak for absorbance ratio calculations. The ratio of absorbance intensity was calculated by using equation (4.1).

$$\text{Absorbance ratio} = \frac{(\text{Absorbance intensity of characteristic peak})}{(\text{Absorbance intensity of a reference peak})} \quad (4.1)$$

The intensity ratios of $A_{3330}/A_{721}$, $A_{1712}/A_{721}$ and $A_{1242}/A_{721}$ are shown in Table 4.9. In regards to the Ar-plasma treated sample, the absorbance ratio is very slightly decreased, as compared to other plasma treated and control samples. In the case of $O_2$ and air plasma treatment, the normalized peak intensity ratio is better than that of the untreated sample at the peaks of 3330 (-OH), 1742 (-C=O) and 1242 (-C-O) $cm^{-1}$. This intensity ratio may indicate the formation of new hydrophilic groups on the surface of fabric during $O_2$ and air plasma treatments [223, 224]. In general, the absorbance ratios measured by ATR-FTIR have not revealed a clear difference among samples. As is already known, the plasma treatment only changes the outermost surface layers of a textile substrate at the depth of a few nanometer ranges (5–50 nm). However, the ATR-FTIR technique is very limited in detecting a thin layer in a nano-level. In order to get more accurate surface chemistry of plasma treated samples, more sophisticated techniques are required [8, 17].

**Table 4.9:** Absorbance groups and their peaks found in the ATR-FTIR spectrum of plasma treated and untreated P/C blend fabric.

| Sample | $A_{3330}/A_{721}$ | $A_{1712}/A_{721}$ | $A_{1242}/A_{721}$ |
|---|---|---|---|
| Untreated P/C | 0.273 | 0.62 | 0.706 |
| $O_2$-plasma treated | 0.386 | 0.622 | 0.708 |
| Ar-plasma treated | 0.304 | 0.617 | 0.695 |
| Air-plasma treated | 0.719 | 0.647 | 0.717 |

Peak intensity at 721 $cm^{-1}$ for C-H stretching was used for normalization.

On the other hand, the ATR-FTIR spectra of plasma coated with $O_2$/HMDSO, $N_2$/HMDSO, $H_2O$/HMDSO and HMDSO, and untreated P/C blend fabric are shown in Figure 4.21. The characteristic peaks and functional groups of untreated P/C fabric were presented along with $O_2$, Ar and air plasma treated fabric in Table 4.8. In this section, more attention is given to ATR-FTIR spectrums of plasma polymerized coating. In Figure 4.21, absorbance bands at 780–980 $cm^{-1}$ assigned to Si-$CH_3$ symmetric bending and Si-$CH_2$ asymmetric stretching; 1000-1200 $cm^{-1}$ assigned to Si-O-Si bending vibration and Si-O-C symmetric stretching; 1240-1270 $cm^{-1}$ assigned to $CH_2$, $CH_3$ and Si-$(CH_3)_x$ symmetric deformation, and 2880-3000 $cm^{-1}$ assigned to C-H stretching from $CH_2$, and $CH_3$ in all plasma polymerized samples [59, 206, 225–227], which are absent in the untreated P/C fabric.

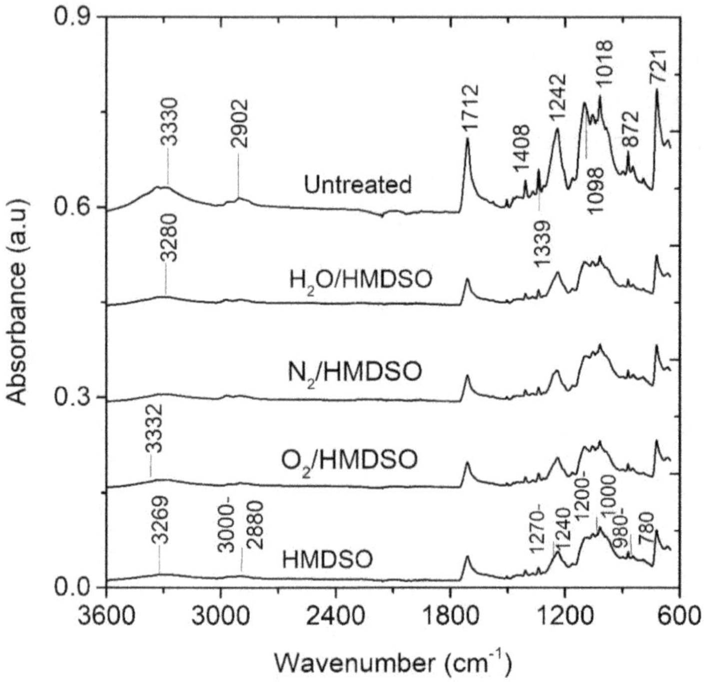


**Figure 4.21.** ATR-FTIR spectra of plasma coated and untreated P/C blend fabric.

As indicated, Si–$CH_3$ groups found in the deposited layer formed by admixture of HMDSO plasma polymerization. In particular, absorbance peaks at 2902 and 2970 cm-1 associated with $CH_3$ symmetric and asymmetric stretching, respectively [228–230]. In addition, C-H stretching (alkyne) and O-H stretching (alcohol) were observed at the characteristic peaks of 3269-3332 and 2700-3200 $cm^{-1}$, correspondingly. In addition, Si–OH stretching was also found at 984 $cm^{-1}$ in the spectras of $O_2$/HMDSO and $H_2O$/HMDSO treated as well as untreated samples. Asymmetric stretching vibration of –C–N–C– bonds may be formed near to 1160

$cm^{-1}$ after $N_2$/HMDSO plasma coating [199]. In fact, The ATR-FTIR spectrum of only liquid HMDSO monomers revealed Si-O-Si stretching near to 1051 $cm^{-1}$. The symmetric bending of $–CH_3$ in Si-$(CH_3)_3$ assigned to at 1252 $cm^{-1}$, while Si-$CH_3$ stretching were noticed around the characteristic peaks of 839 and 754 $cm^{-1}$. In addition, symmetric and asymmetric stretching of methyl groups were observed near to 2900 and 2958 $cm^{-1}$, respectively [228, 231, 232].

Plasma polymerization of HMDSO molecules are firstly dissociated by electron impact ionization process into $(CH_3)_3$-Si-O-Si-$(CH_3)_2^+$ ions and $CH_3^*$ radicals. Subsequently, $(CH_3)_3$-Si-O-Si-$(CH_3)_2^+$ ions further polymerize with the newly introducing HMDSO gas molecules. Consequently, large activated ion complexes are formed [233]. In particular, different linear and cyclic compounds formed via fragmentation of the precursor and condensation of the same units. The identified species including ethoxytrimethylsilane, tetramethyldisiloxane, pentamethyldisiloxane, and heptamethyltrisiloxane. The occurrence of those species containing the dimethylsiloxane (-$Me_2$SiO-) repeating unit reveals the oligomerization reaction. Therefore, chain propagation takes places on the Si-O backbone, resulting in deposition of polymer by the plasma [228, 233].

ATR-FTIR spectra of plasma polymerized fabrics did not show significant difference among $O_2$/HMDSO, $N_2$/HMDSO, $H_2O$/HMDSO and HMDSO. In order to quantify the influence of plasma polymerisation on the surface chemical changes among them, the absorbance ratio of characteristic peaks was analyzed. The absorbance ratios were calculated by dividing absorbance of the characteristic peaks at 2970, 1018, 872 and 793 $cm^{-1}$ to that of the peak at 721 $cm^{-1}$, as tabulated below in Table 4.10. The absorbance ratios increased for all plasma polymerized spectrums of the fabrics. It indicates the higher concentration of Si-containing groups after polymer depositions. Specially, the absorbance ratios of $A_{2970}/A_{721}$ and $A_{793}/A_{721}$ for the untreated fabric were 0.163 and 0.337, which then increased to 0.214 and 0.504 after treatment with $H_2O$/HMDSO and $N_2$/HMDSO plasmas, respectively. The order of these results may not be correlated to the results of TEGEWA drop tests. However, the hydrophilic properties of plasma polymerized fabric depend on the ratio of water vapor, oxygen and nitrogen to the HMDSO monomer.

**Table 4.10:** The absorbance ratio of the peaks in ATR-FTIR spectra of plasma polymerized and untreated P/C blend fabric.

| Sample | $A_{2970}/A_{721}$ | $A_{1018}/A_{721}$ | $A_{872}/A_{721}$ | $A_{793}/A_{721}$ |
|---:|---:|---:|---:|---:|
| Untreated P/C | 0.163 | 0.950 | 0.534 | 0.337 |
| $H_2O$/HMDSO | 0.214 | 0.975 | 0.567 | 0.410 |
| $N_2$/HMDSO | 0.207 | 1.004 | 0.599 | 0.504 |
| $O_2$/HMDSO | 0.176 | 0.984 | 0.564 | 0.399 |
| HMDSO | 0.189 | 1.039 | 0.634 | 0.486 |

Peak intensity at 721 $cm^{-1}$ for C-H stretching was used for normalization.

# CHAPTER FIVE

# CONCLUSION AND FUTURE WORK

## 5.1 Conclusion

In this dissertation, the effect of low-temperature plasma treatment on the comfort properties of P/C blend fabric has been studied in a low pressure plasma reactor. The plasma treatment was conducted on the samples of 65% polyester and 35% cotton blend fabric using argon, oxygen and air, separately, as well as mixtures of $O_2$/HMDSO, $N_2$/HMDSO, $H_2O$/HMDSO and HMDSO monomer alone as working gases. Special attention was given to oxygen plasma due to its wide application for various textile/polymer materials. Due to this reason, the Taguchi method was used to design, analyze and optimize the oxygen plasma process. Other plasma parameters including working pressure, discharge power, duration of treatment and flow rate were also considered. The various comfort properties of plasma treated and untreated P/C blend fabric have been studied. In particular, thermal comfort such as, water vapor, air permeability, thermal resistance and wickability as well as tactile comfort including hand-feel and electro-physical properties were investigated.

In the present work, the important conclusions obtained from the study are summarized as follows.

As the results showed that the oxygen plasma treatment had a positive effect on the thermal comfort properties of P/C the blend fabric. For instance, the wickability of treated fabrics increased in the range of 43.25 to 60.75% and 37.63 to 57.89% compared to untreated fabric in the warp and weft directions, respectively. The reason is the creation of space between inter-fiber and inter-yarns in addition to the hydrophilicity of the fiber surface. Similarly, water vapor permeability of plasma treated samples were enhanced because of the incorporation of oxygen containing polar groups on the fiber surface. The thermal resistance also decreased at least by 20.16% due to the formation new pores by means of surface cleaning and etching effect. This is a good agreement with the air permeability of plasma treated fabric except for experimental runs of 1 and 7. For the decreasing of air permeability at experimental runs of 1 and 7, different possible reasons may be considered. These are the trapping of air with cotton fiber due to plasma treatment or the uniformity of fabric construction. For those thermal comfort properties, optimum parameter levels and influcial variables were clearly showed by using the Taguchi method. The Taguchi analysis confirmed that the pressure had the greatest effect on the wicking and the air permeability during treatment and the flow rate on water vapor permeability, whereas power showed the greatest influence on the thermal resistance of the fabric.

On the other hand, the effect of air, Ar and $O_2$ plasmas on sensorial comfort properties of P/C blend fabric were studied. After plasma treatment, the surface and volume resistivities of the fabric were reduced. However, the fabric softness, smoothness, stiffness and overall HF properties were slightly affected by plasma treatment, as the TSA results revealed. The $O_2$-plasma had a higher resistivity reduction and lower (4%) hand-feel effects due to its chemical nature, whereas Ar-plasma had a lower resistivity reduction and higher hand-feel effects due to its stronger sputtering effect. Therefore, in the case of hand-feel property, selection of gases used for plasma treatment needs a great care.

The surface characterization of all plasma treated and untreated samples were analyzed. Specially, the surface morphology of plasma treated fiber surfaces were changed as the SEM images confirmed. The SEM images have shown the voids, grooves and nano-pits feature on the plasma treated P/C blend fabric surfaces that correspond to the types of plasma gases. As a result, the plasma treated fabric surface became relatively rougher and stiffer than untreated fabric surfaces. Cold plasma treatment has a good potential to modify the clothing comfort properties by altering the outermost surface of the fibers less than a few nanometers in depth; it has also been widely used in the area of textile substrate modification in order to change other surface properties. The outcomes obtained in this study show that the potential of plasma application is suitable to give specific properties to the fabric and to fine-tune them.

Overall, in this dissertation, an attempt has been made to resolve the challenges related to the wet-chemical textile surface modification processes and a new knowledge has been obtained to broaden the application of plasma technology in the textile field. In general, the flexibility of plasma surface modification has opened up many possibilities for using it in textile processing. However, upscaling and transferring a laboratory batch process to a commercial process can also present some technical challenges including the plasma technology itself, textile sector-related issues and specific textile properties. In particular, the optimum parameters must be defined for each plasma process and reactor.

## 5.2 Future work

This study can be considered as the first work that fully focused on the influence of plasma surface modification on the comfort properties of P/C blend fabric using both non-polymer forming and polymer forming plasmas. This could motivate the researchers for further instigations of different blend fabrics by means of various types of plasma gases and monomers. In this dissertation, only 65% polyester and 35% cotton printed blend fabric, ready to make garment, was employed for low pressure plasma treatment. Other scholars can study the blend fabric with various blend ratios at garment level and the effect of plasma on dyed and printed fabric using atmospheric plasmas.

As it is proven in this study, the aging effect is more pronounced when plasma surface modification was performed using reactive (oxygen & air) and inert (Ar) plasma gases. Further studies are recommended using hydrophilic monomers.

Regarding the assessment methods of clothing comfort after plasma treatment, only objective comfort evaluation techniques have been used in this study. The subjective evaluation of clothing comfort can be investigated further in the future. Besides, after plasma treatment, slightly yellowish color in non-polymerising gases at higher operating power and longer treatment time and pale-yellow color in HMDSO monomer plasmas were observed. This also needs further investigation through colorimeters.

Furthermore, concerning surface characterization, additional techniques are required to analyze the surface chemistry, weight loss, thickness of polymer plasma, in the case of precursor, and others. X-ray photoelectron spectroscopy (XPS) is a powerful tool for determining the chemical composition and speciation of material surface for all kinds of plasma treatments. X-Ray Diffraction (XRD) is also required to identify the crystalline phases present in a material and thereby reveal chemical composition information.

## List of publications


[1] **B. B. Yilma**, J. F. Luebben, and M. G. Tadesse, Effect of Plasma Surface Modification on Comfort Properties of Polyester / Cotton Blend Fabric, *Materials Research*, vol. 24, no. 3, pp. 1–11, 2021, doi: https://doi.org/10.1590/1980-5373-MR-2021-0021.

[2] **B. B. Yilma**, J. F. Luebben, and G. Nalankilli, Cold plasma treatment in wet chemical textile processing, *Fibers and Textiles in Eastern Europe*, vol. 28, no. 6, pp. 118–126, 2020, doi: 10.5604/01.3001.0014.3807.

[3] **B. B. Yilma**, J. F. Luebben, and G. Nalankilli, The Effect of Air, Ar and $O_2$ Plasmas on the Electrical Resistivity and Hand-Feel Properties of Polyester/Cotton Blend Fabric, *Fibers*, vol. 8, no. 2, pp. 1–15, 2020, doi:10.3390/fib802001.

## References


[1] Synthetic fibres market analysis, by type (acrylics, polyester, nylon, polyolefin), by application (clothing, home furnishing, automotive, filtration), by region, and segment forecast, 2018-2025, *Grand View Research Market Research Report.* Published date: Nov, 2017, pp. 1–100, [Online]. Available: https://www.grandviewresearch.com/industry-analysis/synthetic-fibers-market.

[2] J. Wu, G. Cai, J. Liu, H. Ge and J. Wang, Eco-friendly surface modification on polyester fabrics by esterase treatment, *Applied Surface Science*, 2014, vol. 295, pp. 150–157.

[3] S. Grishanov, Structure and properties of textile materials, in *Handbook of Textile and Industrial Dyeing*, M. Clark, 1st ed. Oxford, Cambridge, Philadelphia and New Delhi: Woodhead Publishing Limited, 2011, vol. 1, pp. 28–63.

[4] K. Kothari, Thermo-physiological comfort characteristics and blended yarn woven fabrics, *Indian Journal of Fibre & Texti le Research*, 2006, vol. 31, pp. 177–186.

[5] A. Kotb, Predicting Yarn Quality Performance Based on Fibers types and Yarn Structure, *Life Science Journal*, 2012, vol. 9, pp. 37–39.

[6] V. Kerkhof, Innovative by nature - Time to act!, *Lenzing AG.*, 2018, pp. 1–26, [Online]. Available: www.lenzing.com.

[7] F. Sarker, R. K. Prasad, M. R. Howlader and N. Abir, Scope of Dyeing Polyester Cotton (PC) Blended Fabric in Single Bath Process for Water, Energy and Time Saving, *IOSR Journal of Polymer and Textile Engineering*, 2015, vol. 2, no. 3, pp. 12–16.

[8] K. Kale, S. Palaskar, P. J. Hauser and A. El-shafei, Atmospheric pressure glow discharge of helium-oxygen plasma treatment on polyester/cotton blended fabric, *Indian Journal of Fibre & Textile Research*, 2011, vol. 36, pp. 137–144.

[9] K. Fiseha, D. Das and N. K. Palaniswamy, Effect of Blend Ratio of Deep-Grooved Polyester/Cotton Fibers on Mechanical and Comfort Properties of Woven Fabrics, *Fibers and Polymers*, 2019, vol. 20, pp. 2215–2221.

[10] N. Bhat and Y. Benjamin, Surface Resistivity Behavior of Plasma Treated and Plasma Grafted Cotton and Polyester Fabrics, *Textile Research Journal*, 1999, vol. 69, pp. 38–42.

[11] H. Henry, Radioactive Static Eliminators for the Textile Industry, in *Radioisotope Techniques, vol. II*, London: M.H. Stationary Office, 1952, pp. 150–161.

[12] P. Hersh, *Surface Characteristics of Fibers and Textiles, Pt.1*. New York: Marcel Dekker, 1975.

[13] M. Sampath, M. Senthilkumar and G. Nalankilli, Effect of filament fineness on comfort characteristics of moisture management finished polyester knitted fabrics, *Journal of Industrial Textiles*, 2011, vol. 41, pp. 160-173 .

[14] G. Govindachetty, P. Sidhan and C. Venkatraman, Thermo-physiological Comfort Properties of Polyester and Polyester / Acrylic blended Synthetic Fabrics treated with Herbal Finishes, *Journal of Engineered Fibers and Fabrics*, 2014, vol. 9, pp. 115-119.

[15] H. Tokura and T. Midorikawa-Tsurutani, Effects of Hygroscopically Treated Polyester Blouses on Sweating Rates of Sedentary Women at 33°C, *Textile Research Journal*,1985, vol. 55, pp. 178–180.

[16] R. Mather, Surface modification of textiles by plasma treatments, in *Surface modification of textiles*, ed., Q. Wei, 1st ed. New York, USA: Woodhead Publishing Limited, 2009, pp. 296–314.

[17] K. Samanta, M. Jassal and K. Agrawal, Antistatic effect of atmospheric pressure glow discharge cold plasma treatment on textile substrates, *Fibers and Polymers*, 2010, vol. 11, pp. 431–437.

[18] M. Cullough, M. Francis, P. Hall and J. David, Nonlinear Aromatic Polyamide Fiber or Diber Assembly and Method of Preparation, Patent number:

4869951, 1989.

[19] T. Shyr, C. Lien and A. Lin, Coexisting antistatic and water-repellent properties of polyester fabric, *Textile Research Journal*, 2011, vol. 81, pp. 254–263.

[20] X. Yuan, K. Jayaraman and D. Bhattacharyya, Effects of plasma treatment in enhancing the performance of woodfibre-polypropylene composites, *Composites Part A: Applied Science and Manufacturing*, 2004, vol. 35, pp. 1363–1374.

[21] T. Herbert, Atmospheric-pressure cold plasma processing technology, in *Plasma technologies for textiles*, R. Shishoo, 1st ed. Cambridge: Woodhead Publishing Limited in association with The Textile Institut, 2007, pp. 79–128.

[22] A. Jelil, A review of low-temperature plasma treatment of textile materials, *Journal of Materials Science*, 2015, pp. 1–31.

[23] B. Yilma, J. Luebben and G. Nalankilli, The Effect of Air, Ar and $O_2$ Plasmas on the Electrical Resistivity and Hand-Feel Properties of Polyester/Cotton Blend Fabric, *Fibers*, 2020, vol. 8, pp. 1–15.

[24] C. Chao, Blending Cotton and Polyester Fibers: Effects of Processing Methods on Fiber Distribution and Yarn Properties, Georgia Institute of Technology, 1963.

[25] H. Mounir, Q. Khadijah, H. Hany, S. Ebraheem and A. Mofeda, Influence of Sportswear Fabric Properties on the Health and Performance of Athletes, *Fibres and Textiles in Eastern Europe*, 2012, vol. 20, pp. 82–88.

[26] S. Mehta, An experimental investigation on optimizing liquid repellency of fluorochemical urethane finish and its effect on the physical properties of polyester/cotton blended fabric, *Fibers*, 2020, vol. 8, pp. 1–14.

[27] J. Robert and H. Rutherford, *Introduction to Plasma Physics*, 1st ed. Bristol and Philadelphia: IOP Publishing Ltd, 1995.

[28] B. Yilma, J. Luebben and G. Nalankilli, Cold plasma treatment in wet

chemical textile processing, *Fibres and Textiles in Eastern Europe*, 2020, vol. 28, pp. 118–126.

[29] K. Samanta, M. Jassal and K. Agrawal, Atmospheric pressure glow discharge plasma and its applications in textile, *Indian Journal of Fibre and Textile Research*, 2006, vol. 31, pp. 83–98.

[30] R. Ramkrishna, R. Mukesh and M. Subroto, Basics of plasma and its industrial applications in textiles, in *Plasma technologies for textile and appare*, P. B. Nema, S. K. & Jhala, 1st ed. New Delhi: Woodhead Publishing India Pvt. Ltd, 2015, pp. 1–27.

[31] R. Samanta, S. Basak, and K. Chattopadhyay, Environment-Friendly Textile Processing Using Plasma and UV Treatment, in *Textile Science and Clothing Technology Methods*, S. M. Subramanian, 1st ed. Singapore, Heidelberg, New York, Dordrecht, London: Springer Science+Business Media Singapore, 2014, pp. 161–200.

[32] H. Dave, L. Ledwani and K. Nema, Nonthermal plasma: A promising green technology to improve environmental performance of textile industries, in *The Impact and Prospects of Green Chemistry for Textile Technology*, Gandhinagar, India: Elsevier Ltd., 2018, pp. 199–249.

[33] C. Tomasino, J. Cuomo and B. Smith, Plasma Treatments of Textiles, *Journal of Coated Fabrics*, 1996, vol. 25, pp. 115–127.

[34] M. Perucca, Introduction to Plasma Technology for Surface Functionalization, in *Plasma Technology for Hyperfunctional Surfaces. Food, Biomedical and Textile Applications.*, ed., G. B. Hubert Rauscher, Massimo Perucca, 1st ed. Weinheim: WILEY-VCH Verlag GmbH & Co. KGaA, 2010, pp. 1–32.

[35] C. Kan and M. Yuen, Textile Modification with Plasma Treatment, *Research Journal in Textile and Apparel*, 2006, vol. 10, pp. 49–64.

[36] A. Fridman, *Plasma Chemistry*. New York, USA: Cambridge University Press, 2008.

[37] H. Conrads and M. Schmidt, Plasma generation and plasma sources, *Plasma Sources Science and Technology*, 2000, vol. 9, pp. 441–454.

[38] M. Yuen and C. Kan, Influence of low temperature plasma treatment on the properties of ink-jet printed cotton fabric, *Fibers and Polymers*, 2007, vol. 8, pp. 168–173.

[39] Thierry, Plasma Frequency. https://www.thierry-corp.com/plasma-knowledgebase/plasma-frequencies (accessed Dec. 17, 2020).

[40] J. Harry, *Introduction to Plasma Technology Science, Engineering and Applications*. Weinheim: WILEY-VCH Verlag & Co. KGaA, 2010.

[41] U. Riemann, Plasma and sheath, *Plasma Sources Science and Technology*, 2009, vol. 18, pp. 1-11.

[42] K. Nema and B. Jhala, *Plasma technologies for textile and apparel*. New Delhi: Woodhead Publishing India Pvt. Ltd, 2015.

[43] C. Kan, *A Novel Green Treatment for Textiles: Plasma Treatment as a Sustainable Technology*, 1st ed. Boca Raton, London & New York: CRC Press Taylor & Francis Group, 2014.

[44] M. Goddard and H. Hotchkiss, Polymer surface modification for the attachment of bioactive compounds, *Progress in Polymer Science*, 2007, vol. 32, pp. 698–725.

[45] C. Wang, M. Du and Y. Qiu, Uniformity and penetration of surface modification effects of atmospheric pressure plasma jet into textile materials, *Advanced Materials Research*, 2011, vol. 331, pp. 356–359.

[46] D. Sun, Surface Modification of Natural Fibers Using Plasma Treatment, in *Biodegradable Green Composites*, S. Kalia, 1st ed. Hoboken: John Wiley & Sons, 2016, pp. 18–39.

[47] R. Mather, *Surface modification of textiles by plasma treatments*. Woodhead Publishing Limited, 2009.

[48] Z. Peršin, A. Vesel, S. Kleinschek and M. Mozetič, Characterisation of

surface properties of chemical and plasma treated regenerated cellulose fabric, *Textile Research Journal*, 2012, vol. 82, pp. 2078–2089.

[49] A. Demir, Atmospheric Plasma Advantages for Mohair Fibers in Textile Applications, *Fibers and Polymers*, 2010, vol. 11, pp. 580–585.

[50] R. Morent and N. De Geyter, Improved textile functionality through surface modifi cations, in *Functional Textiles for Improved Performance, Protection and Health*, N. P. and G. Sun, 1st ed. Oxford, Cambridge, Philadelphia and New Delhi: Woodhead Publishing Limited in association with The Textile Institute, 2011, pp. 1–528.

[51] K. Chinta, M. Landage and M. Sathish, Plasma Technology & its Application in Textile Wet Processing, *International Journal of Engineering Research & Technology*, 2012, vol. 1, pp. 1–18.

[52] R. Morent and N. De Geyter, Improved textile functionality through surface modification, in *Functional Textiles for Improved Performance, Protection and Health*, ed., G. Pan, & N. Sun, 1st ed. Cambridge, Philadelphia & New Delhi: Woodhead Publishing Limited in association with The Textile Institute, 2011, pp. 3–24.

[53] C. Kan, *A Novel Green Treatment for Textiles. Plasma Treatment as a Sustainable Technology*, 1st ed. Boca Raton, London & New York: CRC Press Taylor & Francis Group, 2015.

[54] L. García et al., Cell proliferation of HaCaT keratinocytes on collagen films modified by argon plasma treatment, *Molecules*, 2010, vol. 15, pp. 2845–2856.

[55] M. Chan, M. Ko and H. Hiraoka, Polymer surface modification by plasmas and photons, *Surface Science Reports*, 1996, vol. 24, pp. 1–54.

[56] D. D'Angelo, Plasma-surface Interaction, in *Plasma Technology for Hyperfunctional Surfaces; Food, Biomedical and Textile Applications*, H. Rauscher, M. Perucca and G. Buyle, 1st ed. Weinheim: WILEY-VCH Verlag GmbH & Co. KGaA, 2010, pp. 63–77.

[57] H. Kale and N. Desaia, Atmospheric pressure plasma treatment of textiles using non-polymerising gases, *Indian Journal of Fibre and Textile Research*, 2011, vol. 36, pp. 289–299.

[58] Y. Wang, The uniform Si-O coating on cotton fibers by an atmospheric pressure plasma treatment, *Journal of Macromolecular Science, Part B: Physics*, 2011, vol. 50, pp. 1739–1746.

[59] C. Vautrin-Ul, F. Roux, C. Boisse-Laporte, L. Pastol and A. Chausse, Hexamethyldisiloxane (HMDSO)-plasma-polymerised coatings as primer for iron corrosion protection: Influence of RF bias, *Journal of Materials Chemistry*, 2002, vol. 12, pp. 2318–2324.

[60] R. Shishoo, Plasma technologies for textiles. Cambridge: Woodhead Publishing Limited in association with The Textile Institute, 2007.

[61] R. Rane, M. Ranjan and S. Mukherjee, Basics of plasma and its industrial applications in textiles, in *Plasma technologies for textile and apparel*, ed., K. Nema, 1st ed. New Delhi: Woodhead Publishing India Pvt. Ltd, 2015, pp. 1–27.

[62] R. Roth, Industrial Plasma Engineering, *Industrial Plasma Engineering*, 2001, doi: 10.1201/9781420034127.

[63] G. Buyle, Nanoscale finishing of textiles via plasma treatment, *Materials Technology*, 2009, vol. 24, pp. 46–51.

[64] R. Morent, N. De Geyter, J. Verschuren, K. De Clerck, P. Kiekens and C. Leys, Non-thermal plasma treatment of textiles, *Surface and Coatings Technology*, 2008, vol. 202, pp. 3427–3449.

[65] R. Morent, N. De Geyter, J.Verschuren, D. Clerck, P. Kiekens, C. Leys, Non-thermal plasma treatment of textiles, *Surface and Coatings Technology*, 2008, vol. 202, pp. 3427–3449.

[66] S. Shahidi and M. Ghoranneviss, Sterilization of cotton fabrics using plasma treatment, *Plasma Science and Technology*, 2013, vol. 15, pp. 1031–1033.

[67] V. Ilić et al., Antifungal efficiency of corona pretreated polyester and polyamide fabrics loaded with Ag nanoparticles, *Journal of Materials Science*, 2009, vol. 44, pp. 3983–3990.

[68] P. Malshe, M. Mazloumpour, A. El-Shafei and P. Hauser, Functional military textile: Plasma-induced graft polymerization of dadmac for antimic robial treatment on nylon-cotton blend fabric, *Plasma Chemistry and Plasma Processing*, 2012, vol. 32, pp. 833–843.

[69] S. Müller, R. Zahn, T. Koburger and K. Weltmann, Smell reduction and disinfection of textile materials by dielectric barrier discharges, 2010, vol. 2, pp. 1044–1048.

[70] H. Xu et al., Influence of absorbed moisture on antifelting property of wool treated with atmospheric pressure plasma, *Journal of Applied Polymer*, 2009, vol. 113, pp. 3687–3692.

[71] S. Shahidi, A. Rashidi, M. Ghoranneviss, A. Anvari and J. Wiener, Plasma effects on anti-felting properties of wool fabrics, *Surface and Coatings Technology*, 2010, vol. 205, pp. 349–354.

[72] C. Kan and M. Yuen, Static properties and moisture content properties of polyester fabrics modified by plasma treatment and chemical finishing, *Nuclear Instruments and Methods in Physics*, 2008, vol. 266, pp. 127–132.

[73] A. Rashidi, H. Moussavipourgharbi, M. Mirjalili, M. Ghoranneviss, Effect of low-temperature plasma treatment on surface modification of cotton and polyester fabrics, *Indian Journal of Fibre and Textile Research*, 2004, vol. 29, pp. 74–78.

[74] Y. Liu, Y. Xiong and D. Lu, Surface characteristics and antistatic mechanism of plasma-treated acrylic fibers, *Applied Surface Science*, 2006, vol. 252, pp. 2960–2966.

[75] H. Yoon and A. Buckley, Improved Comfort Polyester, Part I: Transport Properties and Thermal Comfort of Polyester/Cotton Blend Fabrics, *Textile Research Journal*, 1984, vol. 54, p. 289.

[76] K. Slater, Comfort Properties of Textiles, *Textile Progress,* 1977, vol. 9, pp. 1–42,.

[77] K. Slater, The Assessment of Comfort, *Journal of the Textile Institute*, 1986, vol. 77, pp. 157–171.

[78] F. Rombaldoni, R. Mossotti, A. Montarsolo, R. Demichelis, R. Innocenti and G. Mazzuchetti, The effects of HMDSO plasma polymerization on physical , low-stress mechanical and surface properties of wool fabrics, *AUTEX Research Journal*, 2008, vol. 8, pp. 77–83.

[79] A. Ferri, F. Rombaldoni, G. Mazzuchetti, G. Rovero and S. Sicardi, Thermal Properties of Wool Fabrics Treated in Atmospheric Pressure Post-Discharge Plasma Equipment, *Journal of Engineered Fibers and Fabrics*, 2012, vol. 7.

[80] S. Najar, X. Wang and M. Naebe, The effect of plasma treatment and tightness factor on the low-stress mechanical properties of single jersey knitted wool fabrics, *Textile Research Journal*, 2018, vol. 88, pp. 499–509.

[81] N. Yaman, E. Özdogan and N. Seventekin, Improving physical properties of polyamide fibers by using atmospheric plasma treatments, *Tekstil ve Konfeksiyon*, 2012, vol. 2, pp. 102–105.

[82] N. Yaman, E. Özdoğan and N. Seventekin, Evaluation of some of the physical properties of atmospheric plasma treated polypropylene fabric, *Journal of the Textile Institute*, 2010, vol. 101, pp. 746–752.

[83] S. Sundaresan, K. Thangamani and A. Arunraj, Detailed Study on the Comfort Properties of Zari Fabric, *Journal of Textile Engineering and Fashion Technology*, 2017, vol. 1, pp. 1–8.

[84] C. Prakash, G. Ramakrishnan, S. Chinnadurai, S. Vignesh and M. Senthilkumar, Effect of plasma treatment on air and water-vapor permeability of bamboo knitted fabric, *International Journal of Thermophysics*, 2013, vol. 34, pp. 2173–2182.

[85] J. Venkatesh, K. N. N. Gowda and V. Subramanium, Effect of Air Plasma Treatment on Comfort Properties of Bamboo Fabric, 2013, vol. 2, no. 12,

pp. 2226–2229.

[86] A. Jebastin Rajwin and C. Prakash, Effect of Modified Yarn Path Ring Spinning on Thermal Comfort Properties of Cotton Fabrics after Plasma Treatment, *Journal of Natural Fibers*, 2019, pp. 1–12, DOI: 10.1080/15440478.2019.1650157 .

[87] A. Jebastin Rajwin and C. Prakash, A study on the effect of plasma treatment on thermal comfort properties of cotton fabric, *Journal of Testing and Evaluation*, 2018, pp. 1–10, https://doi.org/10.1520/ JTE20170156.

[88] P. Senthilkumar and T. Karthik, Effect of Argon Plasma Treatment Variables on Wettability and Antibacterial Properties of Polyester Fabrics, *Journal of The Institution of Engineers (India): Series E*, 2016, vol. 97, pp. 19–29.

[89] Y. Li, The science of clothing comfort, *Textile Progress*, 2001, vol. 31, pp. 1–135.

[90] L. Barker, From fabric hand to thermal comfort : the evolving role of objective measurements in explaining human comfort response to textiles, *International Journal of Clothing Science and Technology*, 2002, vol. 14, pp. 181–200.

[91] R. Nielsen, Work clothing, in *International Journal of Industrial Ergonomics*, vol. 7, 1991, pp. 77–85.

[92] T. Bartels, Physiological comfort of sportswear, in *Textiles in Sport*, Cambridge, The Textile Institute: CRC Press, Woodhead Publishing Limited, 2005, pp. 177–203.

[93] Apurba Das and R. Alagirusamy, *Science in Clothing Comfort*, 1st ed. New Delhi: Woodhead Publishing, 2010.

[94] Y. Hu, Y. Li and W. Yeung, Mechanical tactile properties, in *Clothing Biosensory Engineering*, 1st ed., Cambridge, England and New york, USA: Woodhead Publishing Ltd, 2006, pp. 261–284.

[95] M. Dhinakaran, S. Sundaresan and B. S. Dasaradan , Comfort properties of

textiles, *Indian Textile journal*, 2007, vol. 32.

[96] D. Atalie, K. Gideon, A. Ferede, P. Tesinova, and I. Lenfeldova, Tactile Comfort and Low-Stress Mechanical Properties of Half-Bleached Knitted Fabrics Made from Cotton Yarns with Different Parameters, *Journal of Natural Fibers*, 2019, pp. 1–13.

[97] E. Kamalha, Y. Zeng, I. Mwasiagi and S. Kyatuheire, The Comfort Dimension; a Review of Perception in Clothing, *Journal of Sensory Studies*, 2013, vol. 28, pp. 423–444.

[98] R. Liu, T. Lao, L. Kwok and Y. Li, Effects of Graduated Compression Stockings with Different Pressure Profiles on Lower-Limb Venous Structures and Haemodynamics, *Advances in Therapy*, 2008, vol. 25, pp. 465–478.

[99] R. Liu and T. Little, The 5Ps Model to Optimize Compression Athletic Wear Comfort in Sports, *Journal of Fiber Bioengineering and Informatics*, 2009, vol. 2, pp. 41–51.

[100] P. Rajan, G. Ramakrishnan, S. Sundaresan and P. Kandhavadivu, The in fl uence of fabric parameter on low-stress mechanical properties of polyester warp-knitted spacer fabric, *International Journal of Fashion Design, Technology and Education*, 2016, pp. 1–9.

[101] P. Bishop, Fabrics : Sensory and Mechanical Properties, *Textile Progress*, 1996, vol. 26, pp. 1–62.

[102] N. Pan, Auantification and Evaluation of Human Tactile, *International Journal of Design and Nature*, 2007, vol. 1, pp. 48–60.

[103] M. Behery, *Effect of mechanical and physical properties on fabric hand*, 1st ed. Cambridge, England: Woodhead Publishing, 2005.

[104] M. Mitsuo, Fabric handle and its basic mechanical properties, *Journal of Textile Engineering*, 2006, vol. 52, pp. 1–8.

[105] L. Hes, Non-destructive determination of comfort parameters during marketing of functional garments and clothing, *Indian Journal of Fibre and*

*Textile Research*, , 2008, vol. 33, pp. 239–245.

[106] K. Behera, R. Mishra, G. Singh and P. Saighal, Comfort and handle behaviour of linen blended fabrics, *AUTEX Research Journal*, 2007, vol. 7, pp. 33–47.

[107] E. Bertaux, M. Lewandowski and S. Derler, Relationship between Friction and Tactile Properties for Woven and Knitted Fabrics Research Method and Procedures, *Textile Research Journal*, 2007, vol. 77, pp. 387–396.

[108] O. Kim and L. Slaten, Objective Evaluation of Fabric Hand Part I: Relationships of Fabric Hand by the Extraction Method and Related Physical and Surface Properties and FAST techniques less suitable for industrial application, *Textile Research Journal*, 1999, vol. 69, pp. 59–67.

[109] V. Sülar and A. Okur, Sensory evaluation methods for tactile properties of fabrics, *Journal of Sensory Studies*, 2007, vol. 22, pp. 1–16.

[110] P. Saville, *Physical testing of textiles*, 1st ed. Cambridge, England: Woodhead Publishing Ltd, 1999.

[111] I. Frydrych and G. Dziworska, Comparative Analysis of the Thermal Insulation Properties of Fabrics Made of Natural and Man-Made Cellulose Fibres, *Fibres and Textiles in Eastern Europe*, 2002, vol. 10, pp. 40–44.

[112] A. Asanovic, D. Cerovic, V. Mihailovic, M. Kostic and M. Reljic, Quality of clothing fabrics in terms of their comfort properties, *Indian Journal of Fibre and Textile Research*, 2015, vol. 40, pp. 363–372.

[113] K. Jasti, M. Seyam, W. Oxenham and T. Theyson, A novel approach to investigating frictional electrification and charge decay on woven textile fabrics treated with ionic antistatic and hydrophilic surface finishes, *Journal of the Textile Institute*, 2019, vol. 110, pp. 338–348.

[114] W. Ballou, Static Electricity in Textiles, *Textile Research Journal*, 1954, vol. 24, pp. 146–155.

[115] D. Wilson, The Electrical Resistance of Textile materials as a Measure of

their Anti-static Properties, *Journal of the Textile Institute Transactions*, 1963, vol. 54, pp. 97–105.

[116] A. Gonzalez,Electrostatic protection, in *Textiles for protection*, A. Scott, 1st ed. Cambridge, England and New york, USA: Woodhead Publishing Limited and CRC Press LLC, 2005, pp. 503–528.

[117] X. Zhang, *Antistatic and conductive textiles*. Woodhead Publishing Limited, 2011.

[118] T. Malik and K. Sinha, Clothing comfort : A key parameter in clothing," *Physical Therapy Journal*, 2012, pp. 55–57.

[119] C. Parsons, *Human Thermal Environments: The effects of hot, moderate, and cold environments on human health, comfort and performance*, 2nd ed. London and New York: Taylor & Francis Inc, 2003.

[120] A. Angelova, *Textiles and Human Thermophysiological Comfort in the Indoor Environment*, 1st ed. Boca Raton London New york: Taylor & Francis Group, 2016.

[121] Y. Li, Perceptions of temperature, moisture and comfort in clothing during environmental transients, *Ergonomics*, 2005, vol. 48, pp. 234–248.

[122] B. Das, A. Das, K. Kothari, R. Fanguiero and M. Araújo, Moisture transmission through textiles. Part II: Evaluation methods and mathematical modelling, *Autex Research Journal*, 2007, vol. 7, pp. 194–216.

[123] X. Qian and J. Fan, Prediction of clothing thermal insulation and moisture vapour resistance of the clothed body walking in wind, *Annals of Occupational Hygiene*, 2006, vol. 50, pp. 833–842.

[124] L. Hes and C. Loghin, Heat, Moisture and Air Transfer Properties of Selected Woven Fabrics in Wet State, *Journal of Fiber Bioengineering and Informatics*, 2009, vol. 2, pp. 141–149.

[125] C. Mitchell, What is Neurophysiology ?" https://www.wisegeek.com/what-is-neurophysiology.htm (accessed Sep. 08, 2020).

[126] L. Hes and J. William, Laboratory measurement of thermo-physiological comfort, in *Improving comfort in clothing*, Guowen Song, 1st ed. Philadelphia, New Delhi, Oxford, Cambridge: Woodhead Publishing Limited, 2011, pp. 114–137.

[127] A. Iggo, Sensory Receptors, Cutaneous, in *Sensory Systems II: Senses Other than Vision*, 2nd ed., M. Wolfe, 1st ed. Basel: A Pro Scientia Viva Title, 1988, pp. 109–110.

[128] I. S. 7730 The International Organization for Standardization, Ergonomics of the thermal environment – Analytical determination and interpretation of thermal comfort using calculation of the PMV and PPD indices and local thermal comfort criteria (ISO 7730). Geneva, 2005.

[129] P. Sulistiawan, S. Wandri, E. Fauzan and N. Bahari, Thermal Comfort Investigation on Holy Mosque in Bandung Case Study : Lautze 2 Mosque Bandung, *Journal of Architectural Research and Education*, 2020, vol. 2, pp. 82–89.

[130] M. Kosaka, M. Yamane, R. Ogai, T. Kato, N. Ohnishi and E. Simon, Human body temperature regulation in extremely stressful environment : epidemiology and pathophysiology of heat stroke, *Journal of Thermal Biology*, 2004, vol. 29, pp. 495–501.

[131] I. Holmer, Protection against cold, in *Textiles in Sport*, 1st ed., R. Shishoo, Ed. Cambridge, The Textile Institute: CRC Press, Woodhead Publishing Limited, 2005, pp. 262–286.

[132] B. Baker, Physiology of sweat gland function: The roles of sweating and sweat composition in human health, *Temperature*, 2019, vol. 6, pp. 1–49.

[133] L. Lao, D. Shou, Y. Wu and J. Fan, Skin-like' fabric for personal moisture management, *Science Advances*, 2020, vol. 6, pp. 1–11.

[134] A. Marolleaau, F. Salaun, D. Dupont, H. Gidik and S. Ducept, Influence of Textile Physical Properties and Thermo-Hydric Behaviour on Comfort, *Journal of Ergonomics*, 2017, vol. 07.

[135] Q. Zhu and Y. Li, Effects of pore size distribution and fiber diameter on the coupled heat and liquid moisture transfer in porous textiles, *International Journal of Heat and Mass Transfer*, 2003, vol. 46, pp. 5099–5111.

[136] N. Suprun, Dynamics of moisture vapour and liquid water transfer through composite textile structures, *The Eletronic Library*, 2003, vol. 34, pp. 1–5.

[137] C. Phillip W. Gibson, Modeling Convection/diffusion Processes in Porous Textiles with Inclusion of Humidity-dependent Air Permeability, *Int. Comm. Heat Mass Transfer*, 1997, vol. 24, pp. 709–724.

[138] O. Fanger, Thermal environment-Human requirements, *Sulzer Technical Rev*, 1985, vol. 67, pp. 3–6.

[139] L. Yi et al., P-smart-a virtual system for clothing thermal functional design, *Computer Aided Design*, 2006, vol. 38, pp. 726–739.

[140] Y. Li and V. Holcombe, Mathematical Simulation of Heat and Moisture Transfer in a Human-Clothing-Environment System, *Applied Mechanics and Materials*, 1998, vol. 68, pp. 389–397.

[141] L. Fourt, *Clothing Comfort and function*. New york, NY: Marcel Dekker Inc., 1970.

[142] W. Dent, Transient comfort phenomena due to sweating, *Textile Research Journal*, 2001, vol. 71, pp. 796–806.

[143] B. McNeill and C. Parsons, Appropriateness of international heat stress standards for use in tropical agricultural environments, *Ergonomics*, 1999, vol. 42, pp. 779–797.

[144] B. de Vasconcelos, M. de Barros, C. Borelli and G. de Vasconcelos, Moisture Management Evaluation in Double Face Knitted Fabrics with Different Kind of Constructions and Fibers, *Journal of Fashion Technology & Textile Engineering*, 2017, vol. S3:009, pp. 1–5.

[145] S. Rengasamy, Improving moisture management in apparel, in *Improving Comfort in Clothing*, 1st ed., G. Song, 1st ed. Cambridge, Philadelphia, New

Delhi: The Textile Institute and Woodhead Publishing, 2011.

[146] G. de Camargo et al., Morphological and Chemical Effects of Plasma Treatment with Oxygen ($O_2$ ) and Sulfur Hexafluoride ($SF_6$ ) on Cellulose Surface, *Materials Research*, 2017, vol. 20, pp. 842–850.

[147] A. Ibrahim, M. Eid, A. Youssef, A. Ameen and M. Salah, Surface modification and smart functionalization of polyester-containing fabrics, *Journal of Industrial Textiles*, 2013, vol. 42, pp. 353–375.

[148] E. Onofrei, M. Rocha and A. Catarino, The influence of knitted fabrics' structure on the thermal and moisture management properties, *Journal of Engineered Fibers and Fabrics*, 2011, vol. 6, pp. 10–22.

[149] H. Özdemir, Thermal comfort properties of clothing fabrics woven with polyester/cotton blend yarns, *Autex Research Journal*, 2017, vol. 17, pp. 135–141.

[150] M. El Messiry, A. El Ouffy and M. Issa, Microcellulose particles for surface modification to enhance moisture management properties of polyester, and polyester/cotton blend fabrics, *Alexandria Engineering Journal*, 2015, vol. 54, pp. 127–140.

[151] J. Yuan, Z. Zhang, M. Yang, F. Guo, X. Men and W. Liu, Surface modification of hybrid-fabric composites with amino silane and polydopamine for enhanced mechanical and tribological behaviors, *Tribiology International*, 2017, vol. 107, pp. 10–17.

[152] M. Kamel, G. Allam, K. El Gabry and M. Helmy, Surface Modification Methods for Improving Dyeability of Acrylic Fabric Using Natural Biopolymer, *Journal of Applied Sciences Research*, 2013, vol. 9, pp. 3520–3529.

[153] G. Li, H. Liu, T. Li and J. Wang, Surface modification and functionalization of silk fibroin fibers/fabric toward high performance applications, *Materials Science and Engineering C*, 2012, vol. 32, pp. 627–636.

[154] H. Lee and S. Song, Effects of Treatments with Two Lipolytic Enzymes on Cotton/Polyester Blend Fabrics, *Journal of the Korean Society of Clothing*

*and Textiles*, 2013, vol. 37, pp. 1107–1116.

[155] S. Shahidi, J. Wiener and M. Ghoranneviss, Surface Modification Methods for Improving the Dyeability of Textile Fabrics, in *Eco-Friendly Textile Dyeing and Finishing*, M. Gunay, 1st ed. IntechOpen, 2013, pp. 31–51.

[156] C. Deogaonkar, Dielectric barrier discharge plasma induced surface modification of polyester/cotton blend fabrics to improve polypyrrole coating adhesion and conductivity, *Journal of the Textile Institute*, pp. 1–9, 2020, doi: 10.1080/00405000.2019.1710905.

[157] S. Kumar and E. Gunasundari, Sustainable Wet Processing-An Alternative Source for Detoxifying Supply Chain in Textiles, in *Textile Science and Clothing Technology*, S.S. Muthu, 1st ed. Singapore: Springer, 2018, pp. 37–60.

[158] A. Ferri, F. Rombaldoni, G. Mazzuchetti, G. Rovero and S. Sicardi, Thermal properties of wool fabrics treated in atmospheric pressure post-discharge plasma equipment, *Journal of Engineered Fibers and Fabrics*, 2012, vol. 7, pp. 75–81.

[159] B. Karaca, A. Demir, E. Özdoğan and Ö. Işmal, Environmentally benign alternatives: Plasma and enzymes to improve moisture management properties of knitted PET fabrics, *Fibers and Polymers*, 2010, vol. 11, pp. 1003–1009.

[160] J. Rajwin and C. Prakash, Effect of Air Plasma Treatment on Thermal Comfort Properties of Woven Fabric, *International Journal of Thermophysics*, 2017, vol. 38.

[161] T. Peirce, The handle of cloth as a measurable quantity, *Journal of the Textile Institute Transactions*, 1930, vol. 21, pp. T377–T416.

[162] S. Kilinc-Balci and Y. Elmogahzy, Testing and analyzing comfort properties of textile materials for the military, in *Military Textiles*, E. Wilusz, Ed. Cambridge: Woodhead Publishing, 2008, pp. 107–36.

[163] I. Holmér, Thermal manikin history and applications, *European Journal of*

*Applied Physiology*, 2004, vol. 92, pp. 614–618.

[164] AATCC, Liquid Moisture Management Properties of Textile Fabrics, *Test Method 195-2011*, 2011, pp. 366–370.

[165] N. Uren and A. Okur, Analysis and improvement of tactile comfort and low-stress mechanical properties of denim fabrics, *Textile Research Journal*, 2019, vol. 0, pp. 1–16.

[166] M. Abu-Rous, E. Liftinger, J. Innerlohinger, B. Malengier and S. Vasile, A new physical method to assess handle properties of fabrics made from wood-based fibers, in *IOP Conference Series: Materials Science and Engineering*, 2017, vol. 254.

[167] M. Rous, B. Malengier, E. Liftinger and J. Inneriohnger, Handfeel of Single Jersey Fabrics as Assessed by a New Physical Method, *Journal of Fashion Technology & Textile Engineering,* 2018, vol. S4.

[168] AATCC Test Method 76, "Electrical Surface Resistivity of Fabrics, AATCC Standard, U.S.A., 2005.

[169] S. Palaskar, H. Kale, S. Nadiger and N. Desai, Dielectric Barrier Discharge Plasma Induced Surface Modification of Polyester/Cotton Blended Fabrics to Impart Water Repellency Using HMDSO, *Journal of Applied Polymer Science*, 2011, vol. 122, pp. 1092–1100.

[170] Diener electronic GmbH + Co. KG, Low-pressure plasma laboratory system (Tetra 30, S. No. 114275). Operating Instructions, *Diener electronic GmbH + Co. KG*. Ebhausen, Germany, pp. 1–102, Jan. 2018.

[171] A. Anders and Y. Yushkov, Low-energy linear oxygen plasma source, *Review of Scientific Instruments,* 2007, vol. 78, .

[172] F. Sensors, Mass flow versus volumetric flow. pp. 1–4, 2018, [Online]. Available:https://www.firstsensor.com/cms/upload/appnotes/AN_Massflow_E_11153.pdf.

[173] J. Legerská and P. Lizák, Evaluation of the specific physiological properties

for the selected assortment of work clothes, *Procedia Engineering*, 2016, vol. 136, pp. 227–232.

[174] Emtec TSA – Textile Softness Analyzer - Techtextil & Texprocess. Available online: https://techtextil-texprocess.messefrankfurt.com › techtextil (accessed on November18, 2019).

[175] ASTM standard D 257, Standard Test Methods for DC Resistance or Conductance of Insulating Materials, West Conshohocken, Pennsylvania, U.S.A., 2007.

[176] D. Schuttlefield and H. Grassian, ATR-FTIR spectroscopy in the undergraduate chemistry laboratory part I: Fundamentals and examples, *Journal of Chemical Education*, 2008, vol. 85, pp. 279–281.

[177] M. Amitava, The Taguchi method, *WIREs Computational Statistics*, 2011, vol. 3, pp. 472–480.

[178] G. Ramakrishnan, C. Prakash and G. Janani, Effect of environmentally friendly plasma treatment on microfibre fabrics to improve some comfort properties, *International Journal of Clothing Science and Technology*, 2018, vol. 30, pp. 29–37.

[179] R. Van Amber, A. Wilson, M. Laing, B. Lowe and B. E. Niven, Thermal and moisture transfer properties of sock fabrics differing in fiber type, yarn, and fabric structure, *Textile Research Journal*, 2015, vol. 85, pp. 1269–1280.

[180] J. Wei, S. Xu, H. Liu, L. Zheng and Y. Qian, Simplified model for predicting fabric thermal resistance according to its microstructural parameters, *Fibres and Textiles in Eastern Europe*, 2015, vol. 23, pp. 57–60.

[181] H. Karahan, E. Özdoğan, A. Demir, H. Ayhan, and N. Seventekin, Effects of Atmospheric Pressure Plasma Treatments on Certain Properties of Cotton Fabrics, *Fibres and Ttextiles in Eastern Europe*, 2009, vol. 17, pp. 19–22.

[182] R. Prabha and N. Vasugi, Evaluating the Comfort Properties of Plasma Treated and Plasma Untreated Cotton Fabrics Finished with Plant Extracts, *International Journal of Innovative Research in Science, Engineering and*

*Technology*, 2016, vol. 5, pp. 20061–20066.

[183] A. Das, K. Kothari and A. Sadachar, Comfort characteristics of fabrics made of compact yarns, *Fibers and Polymers*, 2007, vol. 8, pp. 116–122.

[184] H. Ozdemir, Permeability and wicking properties of modal and Lyocell woven fabrics used for clothing, *Journal of Engineered Fibers and Fabrics*, 2017, vol. 12, pp. 12–21.

[185] M. Mazloumpour, F. Rahmani, N. Ansari, H. Nosrati and A. H. Rezaei, Study of wicking behavior of water on woven fabric using magnetic induction technique, *Journal of the Textile Institute*, 2011, vol. 102, pp. 559–567.

[186] M. Hossain, D. Hegemann, S. Herrmann and P. Chabrecek, Contact angle determination on plasma-treated poly (ethylene terephthalate) fabrics and foils, *Journal of Applied Polymer Science*, 2006, vol. 102, pp. 1452–1458.

[187] N. Mao, Y.Wang and J. Qu, Smoothness and Roughness : Characteristics of Fabric-To- Fabric Self-Friction Properties, *The 90th Textile institute world conference. Textiles: Inseparablefrom the human environment.*, 2016.

[188] M. Naebe, D. Tester, A. Mcgregor and X. Wang, The effect of plasma treatment and loop length on the handle of lightweight jersey fabrics as assessed by the Wool HandleMeter, *Textile Research Journal*, 2015, vol. 85, pp. 1190–1197.

[189] O. Hung and C. Kan, Effect of $CO_2$ laser treatment on the fabric hand of cotton and cotton/polyester blended fabric, *Polymers*, 2017, vol. 9.

[190] S. Vasile, L. Ciesielska-Wróbel and V. Langenhove, Wrinkle recovery of flax fabrics with embedded superelastic shape memory alloys wires, *Fibres and Textiles in Eastern Europe*, 2012, vol. 93, pp. 56–61.

[191] C. Kanand M.Yuen, Influence of Plasma Gas on the Mechanical Properties of Wool Fabric, *IEEE Transactions on Plasma Science*, 2009, vol. 37 pp. 653–658.

[192] M. Murakami, M. Mori, T. Fujimoto and C. Murata, Influence of Plasma

Treatment on Total Hand Value and Adhesion Properties of Bi-Fabrics of Fused Interlining and Wool Fabrics, *International Conference on Kansei Engineering and Emotion Research Influence*, 2014, no. June 11-13, pp. 1031–1038.

[193] B. Yilma, J. Luebben and G. Nalankilli, The Effect of Low Pressure Oxygen Plasma Treatment on Comfort Properties of Polyester / Cotton Blend Fabric, *International Journal on Textile Engineering and Processes*, 2018, vol. 4, pp. 1–10.

[194] M. Chongqi, Z. Shulin and H. Gu, Anti-static charge character of the plasma treated polyester filter fabric, *Journal of Electrostatics*, 2010, vol. 68, pp. 111–115.

[195] N. Bhat and Y. Benjamin, Surface Resistivity Behavior of Plasma Treated and Plasma Grafted Cotton and Polyester Fabrics, *Textile Research Journal*, 1999, vol. 69, pp. 38–42.

[196] A. Wróbel, M. Kryszewski, W. Rakowski, M. Okoniewski and Z. Kubacki, Effect of plasma treatment on surface structure and properties of polyester fabric, *Polymer*,1978, vol. 19, pp. 908–912.

[197] P. Pransilp, M. Pruettiphap, W. Bhanthumnavin, B. Paosawatyanyong and S. Kiatkamjornwong, Surface modification of cotton fabrics by gas plasmas for color strength and adhesion by inkjet ink printing, *Applied Surface Science*, 2016, vol. 364, pp. 208–220.

[198] S. Canbolat, M. Kilinc and D. Kut, The Investigation of the Effects of Plasma Treatment on the Dyeing Properties of Polyester / Viscose Nonwoven Fabrics, *Procedia - Social and Behavioral Sciences*, 2015, vol. 195, pp. 2143–2151.

[199] N. Ibrahim, M. Hashem, M. Eid, R. Refai, M. El-Hossamy and M. Eid, Eco-friendly plasma treatment of linen-containing fabrics, *Journal of the Textile Institute*, 2010, vol. 101, pp. 1035–1049.

[200] W. Tsoi, C. Kan and M. Yuen, Using Ageing Effect for Hydrophobic

Modification, *BioResources*, 2011, vol. 6, pp. 3424–3439.

[201] M. Tsuchida and Z. Osawa, Effect of ageing atmospheres on the changes in surface free energies of oxygen plasma-treated polyethylene films, *Colloid and Polymer Science*, 1994, vol. 272, pp. 770–776.

[202] C. Kan and Y. Lam, The Effect of Plasma Treatment on Water Absorption Properties of Silk Fabrics, *Fibers and Polymers*, 2015, vol. 16, pp. 1705–1714.

[203] A. Malek and I. Holme, The effect of plasma treatment on some properties of cotton, *Iranian Polymer Journal*, 2003, vol. 12, pp. 271–280.

[204] B. Kilic, C. Aksit and M. Mutlu, urface modification and characterization of cotton and polyamide fabrics by plasma polymerization of hexamethyldisilane and hexamethyldisiloxane, *International Journal of Clothing Science and Technology*, 2009, vol. 21, pp. 137–145.

[205] M. Gorjanc, V. Bukošek, M. Gorenšek and A. Vesel, The Influence of Water Vapor Plasma Treatment on Specific Properties of Bleached and Mercerized Cotton Fabric, *Textile Research Journal*, 2010, vol. 80, pp. 557–567.

[206] H. Kale and S. Palaskar, Atmospheric pressure plasma polymerization of hexamethyldisiloxane for imparting water repellency to cotton fabric, *Textile Research Journal*, 2010, vol. 81, pp. 608–620.

[207] G. Rosace, R. Canton and C. Colleoni, Plasma enhanced CVD of SiO x C y H z thin film on different textile fabrics: Influence of exposure time on the abrasion resistance and mechanical properties, *Applied Surface Science*, 2010, vol. 256, pp. 2509–2516.

[208] K. Samanta, M. Jassal and K. Agrawal, Antistatic effect of atmospheric pressure glow discharge cold plasma treatment on textile substrates, *Fibers and Polymers*, 2010, vol. 11, pp. 431–437.

[209] C. Kan, C. Lam, C. Chan and S. Ng, Using atmospheric pressure plasma treatment for treating grey cotton fabric, *Carbohydrate Polymers*, 2014, vol. 102, pp. 167–173.

[210] C. Wang et al., Surface modification of polyester fabric with plasma pretreatment and carbon nanotube coating for antistatic property improvement, *Applied Surface Science*, 2015, vol. 359.

[211] T. Nguyen, T. Kim, K. Vu, T. Hong, N. Thi and H. Manh, The Effect of DBD Plasma Activation Time on the Dyeability of Woven Polyester Fabric with Disperse Dye, *Polymers*, 2021, vol. 13, pp. 1–22.

[212] C. Zhang and K. Fang, Surface modification of polyester fabrics for inkjet printing with atmospheric-pressure air/Ar plasma, *Surface and Coatings Technology*, 2009, vol. 203, pp. 2058–2063.

[213] K. Rani, B. Sarma and A. Sarma, Plasma treatment on cotton fabrics to enhance the adhesion of Reduced Graphene Oxide for electro-conductive properties, *Diamond and Related Materials*, 2018, vol. 84, pp. 77–85.

[214] S. Shahidi, M. Ghoranneviss, J. Wiener, B. Moazzenchi and H. Mortazavi, Effect of hexamethyldisiloxane (HMDSO)/nitrogen plasma polymerisation on the anti felting and dyeability of wool fabric, *Fibres and Textiles in Eastern Europe*, 2014, vol. 22, pp. 116–119.

[215] S. Asadollahi, J. Profili, M. Farzaneh and L. Stafford, Development of Organosilicon-based superhydrophobic coatings through atmospheric pressure plasma polymerization of HMDSO in nitrogen plasma, *Materials*, 2019, vol. 12, pp. 1–21.

[216] B. Kayaoglu, E. Ozturk, F. Guner and T. Uyar, Improving hydrophobicity on polyurethane-based synthetic leather through plasma polymerization for easy care effect, *Journal of Coatings Technology and Research*, 2013, vol. 10, pp. 549–558.

[217] J. Yang, Y. Pu, D. Miao and X. Ning, Fabrication of durably superhydrophobic cotton fabrics by atmospheric pressure plasma treatment with a siloxane precursor, *Polymers*, 2018, vol. 10.

[218] M. Moghaddam et al., Hydrophobisation of wood surfaces by combining liquid flame spray (LFS) and plasma treatment: Dynamic wetting properties,

*Holzforschung*, 2016, vol. 70, pp. 527–537.

[219] L. Levin et al., In-line material analysis of 50nm defects by integration of Energy (EDX) and Wavelength (WDX) Dispersive X-ray analysis, *International Symposium on Semiconductor Manufacturing (ISSM)*, 2008, Tokyo, Japan, pp. 251–254.

[220] C. Kan and Y. Lam, Low stress mechanical properties of plasma-treated cotton fabric subjected to zinc oxide-anti-microbial treatment, *Materials*, 2013, vol. 6, pp. 314–333.

[221] A. Gore and A. Venkataraman, Identification of polyester/cellulosic blends using FT-IR spectrometer, *Indian Journal of Fibre and Textile Research*, 1998, vol. 23, pp. 165–169.

[222] L. da Silva et al., Surface modification of polyester fabric by non-thermal plasma treatment, in *Journal of Physics: Conference Series*, 2012, vol. 406.

[223] C. Kan and C. Ioghim, Atmospheric Pressure Plasma Treatment for Grey Cotton Knitted Fabric, *Polymers*, 2018, vol. 10, pp. 1–16.

[224] D. Caschera et al., Effects of plasma treatments for improving extreme wettability behavior of cotton fabrics, *Cellulose*, 2014, vol. 21, pp. 741–756.

[225] J. Yang, Y. Pu, D. Miao and X. Ning, Fabrication of durably superhydrophobic cotton fabrics by atmospheric pressure plasma treatment with a siloxane precursor, *Polymers*, 2018, vol. 10.

[226] A. Sarmadi, HMDSO-plasma Modification of Polypropylene Fabrics, *Euroupe polymer journal*, 1995, vol. 31, pp. 847–857.

[227] Z. Ziari et al., Chemical and electrical properties of HMDSO plasma coated polyimide," *Vacuum*, 2013, vol. 93, pp. 31–36.

[228] H. Kale, S. Palaskar and N. Bhat, Effect of plasma polymerization of hexamethyldisiloxane on polyester/cotton blended fabrics properties, *AATCC Review*, 2012, vol. 12, no. 2, pp. 53–60.

[229] K. Rani, B. Sarma, and A. Sarma, Plasma treatment on cotton fabrics to

enhance the adhesion of Reduced Graphene Oxide for electro-conductive properties, *Diamond and Related Materials*, 2018, vol. 84, pp. 77–85.

[230] H. Kale and S. Palaskar, Structural studies of plasma polymers obtained in pulsed dielectric barrier discharge of TEOS and HMDSO on nylon 66 fabrics, *Journal of the Textile Institute*, 2012, vol. 103, pp. 1088–1098.

[231] F. Boerio, V. Wagh and R. Dillingham, Vapor-phase silanation of plasma-polymerized silica-like films by γ-aminopropyltriethoxysilane, *Journal of Adhesion*, 2005, vol. 81, pp. 115–142.

[232] J. Schwarz, M. Schmidt and A. Ohl, Syndissertation of plasma-polymerized hexamethyldisiloxane (HMDSO) films by microwave discharge, *Surface and Coatings Technology*, 1998, vol. 98, pp. 859–864.

[233] Q. Jiao, A. DeJoseph and A. Garscadden, Ion chemistries in hexamethyldisiloxane, *Journal of Vacuum Science and Technology A: Vacuum, Surfaces, and Films*, 2005, vol. 23, pp. 1295–1304.

www.ingramcontent.com/pod-product-compliance
Ingram Content Group UK Ltd.
Pitfield, Milton Keynes, MK11 3LW, UK
UKHW022000190726
13853UKWH00004B/1649